About the Authors

Professor Noor Mohammad (*b.* 1942) after obtaining his M. A. Degree in Geography from the University of Gorakhpur with a first class first, joined Aligarh Muslim University, Aligarh in 1965 as a junior Research Fellow. He was awarded the degree of Ph.D. in 1969 and joined Kirori Mal College in the same year. Since 1976 Prof. Mohammad has been associated with the Delhi School of Economics where he has taught Geography to Post-Graduate and M.Phil. students and has supervised research in the field of Agricultural Geography and Applied Geography to M.Phil. and Ph. D. scholars. Prof. Mohammad is actively engaged in research in Applied Geography and Agricultural Geography and has more than hundred research papers to his credit in many scholarly referred scientific journals. He has authored and edited more than two dozen books. His books on *Agricultural Landuse in India, Impact of Caste on Primary Occupations* and *Spatial Information Technology for Resource Management* have been well taken the world over. Prof. Mohammad's books *Perspectives in Agricultural Geography* (in 5 Volumes) and *New Dimensions in Agricultural Geography* (in 8 Volumes) have been well received and have won wide acclaim in Indian and foreign Universities.

Prof. Mohammad has been associated with the Delhi School of Economics, Delhi University as Assistant Professor, Associate Professor, Professor, Head of Department and Co-Coordinator UGC SAP Programme. He is also associated in various positions and capacities with many Institutions and Organisations such as UGC, ICSSR, NCERT, CBSE, NIOS, etc. He has completed many research projects of ICSSR, UGC and NCERT. Prof. Mohammad has been awarded UGC Emeritus Fellowship and has completed the latest research report on *Impact of Sprinkler Irrigation on Agricultural Development in Drought Prone Areas of Haryana.*

Dr. Suresh C. Rai (*b.*1958) having received his M.Sc. and Ph. D. degree from B. H. U. Varanasi. Joined G.B. Pant Institute of Himalayan. Environment and Development at its Sikkim and Northeast Unit. He has joined the Department of Geography, Delhi University as Associate Professor in 2005. His principal areas of Research are Watershed Management, Mountain Hydrology, Ecotourism and Biodiversity Conservation of Himalayas. He has published over 75 research papers in refereed journal and has edited/authored many books.

Agricultural Diversification and Food Security in the Mountain Ecosystem

A Case Study of Uttarakhand

Prof. Noor Mohammad
UGC Emeritus Fellow

Dr. S.C. Rai
Associate Professor

Department of Geography,
Delhi School of Economics,
University of Delhi, Delhi - 110007

CONCEPT PUBLISHING COMPANY PVT. LTD.
NEW DELHI-110059

ISBN-13 : 978-93-5125-048-7

First Published 2014

Published and Printed by

Concept
Since 1974

Concept Publishing Company Pvt. Ltd.
A/15-16, Commercial Block, Mohan Garden
New Delhi-110059 (India)
T: 25351460, 25351794, *F*: 091-11-25357109
E: publishing@conceptpub.com, W: www.conceptpub.com
Editorial Office: H-13, Bali Nagar, New Delhi-110 015, India.

Cataloging in Publication Data—*Courtesy:* D.K. Agencies (P) Ltd. <docinfo@dkagencies.com>

Mohammad, Noor, 1942-
Agricultural diversification and food security in the mountain ecosystem : a case study of Uttarakhand / Noor Mohammad, S.C. Rai.

p. cm.
Includes bibliographical references (p.) and index.
ISBN 9789351250487

1. Agricultural diversification—India—Uttarakhand. 2. Food supply—India—Uttarakhand. I. Rai, Suresh Chand, 1958-, joint author. II. Title.

DDC 338.1095451 23

Preface

The green revolution has brought tremendous development in agriculture specially in case of foodgrains production which has increased more than five times since independence but it has created a number of environmental problems. Therefore, there is growing awareness about the need to substitute the current production system with a more sustainable and integrated system of agricultural production. The objective can be fulfilled only by ushering in a second Green Revolution, where main planks would be diversification of agriculture in favour of high value commodities. The diversification of agriculture may include animal husbandry, vegetable farming, floriculture, horticulture, agro-forestry, piggries, bee farming, sericulture, acquaculture, poultry farming and so on which will not only add to the agricultural production but will also improve the quality of food.

Since there have been a lot of spatio-temporal variations in the induction and consequently achievements of green revolution and its resulting environmental problems, there is a need for in depth research in this direction. Therefore, the state of Uttarakhand has been selected for the present study due to: its heterogeneous geographical environment-physical, techno-economic and socio-cultural which provides huge potential for agricultural development; implementation of various programmes and policies for agricultural development which has brought changes specially in agricultural land use; its huge potential for agricultural diversification due to diversified physiographic conditions leading to biodiversity; and the southern portion of the state has practiced almost all modern agricultural inputs and has started facing its adverse consequences.

The study is based on primary and secondary data. The secondary data have been collected from published and unpublished records of the government and semi-government organizations and also from the NGO's. The primary data have been collected for 538 households chosen from 30 selected villages lying in different physical regions of the state. The data have been processed and analysed using appropriate statistical techniques and have been presented in tabular, diagrammatic and map forms using GIS software.

The study is organized in eight chapters. It opens up with an introductory chapter dealing with conceptual framework, statement of problem, research questions, aim and objectives, hypotheses, study area and overview of literature based on themes. Chapter 2 deals with research design highlighting in detail the procedure of primary data collection, data base and methods of data processing, analysis and presentation. The geographical profile of the study area has been presented in Chapter 3 which includes physical, economic and social setting of the area. In this chapter only those aspects of geographical environment have been taken which are interlinked with one another and have a bearing on the agricultural diversification and food security in the state.

Chapter 4 discusses the spatial pattern of land use in 2005-06 with a comparison with the year 1992-93 and 2003-04 whenever it was necessary. All the five major categories of land use have been systematically analyzed focusing on those aspects which could help in agricultural diversification and food security. The status and role of irrigation has also been critically examined in this chapter.

Chapters 5, 6 and 7 are the core chapters of the study. Chapter 5 presents a dynamic pattern of agricultural land use and various related components. It examines the special agricultural practices prevalent in Uttarakhand and discusses comprehensively the distributional pattern of major crops highlighting their status, and role in food security. It also discusses the intensity of cropping, crop-combination regions and agricultural diversification index using most suitable statistical techniques. Chapter 6 analyses agricultural diversification and food security at district level using secondary data. Besides explaining the concept and indicators of food security, it has elaborated the data base and methods of its processing and analysis. Finally, it has analyzed and explained the distribution of consumption units, their spatial pattern of distribution, carrying capacity of land in calorific and monetary values and Standard Nutrition Units (SNU). At the end it has discussed and explained the spatial pattern of food security in terms of calorific and monetary values.

Chapter 7 is entirely based on primary data collected during the course of fieldwork. It opens up with personal and household characteristic of the respondents and their farming characteristics. It presents a comprehensive account of agricultural diversification including dairy farming, vegetable farming, horticulture, aquaculture, floriculture, poultry farming and apiculture, and highlights their potentiality and role in food security system. In the end it examines the households and village level food security in calorific and monetary value.

Based on the critical analysis of the facts and figure in the previous chapters, summary, conclusion and suggestions have been given in last chapter which derives innovative conclusions and gives pragmatic and practical suggestions.

Delhi
15th March, 2013

Prof. Noor Mohammad
S.C. Rai

Acknowledgements

Let me first of all thank the Almighty, Whose grace and mercy has always been with me and has provided all I needed which I myself was not aware of.

The completion of this project has been possible due to the constant inspiration and unstinting co-operation of a host of people too numerous to acknowledge individually. However, first and the foremost I express my deep sense of indebtedness and gratitude to my *Guru* and mentor late Prof. Mohammad Shafi. Professor Emeritus, Department of Geography, Aligarh Muslim University, Aligarh who has been a source of inspiration and guidance throughout my academic career. Whatever, I have achieved in my academic pursuits, is due to his persistent and sincere able guidance.

I owe a deep sense of gratitude to my late parents who have shaped and made me what I am to-day. It is their soul and spirit which is still inspiring and giving guidance and strength in every walk of my life.

I express my sincere thanks to University Grants Commission especially to its former officiating Secretary Dr. R.K. Chauhan who has generously sanctioned this major project to me with maximum permissible funds.

I extend my thanks to the Head, Department of Geography, Delhi University who has kindly permitted me to use the infrastructural facilities of the department. Special thanks are due to Dr. S.C. Rai - Associate Professor in the same department who has been kind enough to accept the co-investigator ship in the project which is mandatory for retired teachers to get major research project from any funding agency for financial reasons. I would also like to extend my thanks to all the Faculty members of the department who have helped me in one way or the other in the execution of this project.

I have consulted many libraries within and outside Delhi specially the libraries of Delhi University, Jawaharlal Nehru University, Jamia Milla

Islamia, British Council, ICSSR, NCERT, American, Kumaon University (Nainital) and Garhwal University (Srinagar). I extend my heartiest thanks to all the librarians and their staffs who have been kind enough to allow me to use the relevant data and existing valuable literature.

The entire field work and collection of primary and secondary data has been done by project fellows specially Mr. Rajesh Kumar. I place on record my thanks and appreciation for his pain staking job in the study villages where he stayed for considerable period of time and has collected very valuable primary and secondary data. We are also grateful to respondents of all selected villages, village heads, and the officials of districts, tahsils and block development offices. Needless to mention that without their help and co-operation field investigation would have been very difficult. It is the lack of space rather than my sense of gratitude which bars me from naming them individually.

I would like to add the special words of thank to Prof. G.L. Shah, Department of Geography, Kumaon University, Nainital who has been torch bearer in giving minute details of all aspects of Uttarakhand and has provided homely atmosphere during my stay in Nainital. In this continuation the help and assistance provided by Dr. K.K. Sharma, and Dr. S.K. Bandooni is sincerely acknowledged.

I would also like to express my thanks to M/s Arun Thesis for typing the manuscript and bringing it in the present form. Thanks are also due to Dr. Perveen Misra for drawing all the maps and diagrams within a short period of time.

I would like to put on record my sincere thanks and deep sense of appreciation to my wife Mrs. S. Nisa and children-Nuzhat, Naved, Zareena and Javed who have provided a favourable working environment to me for successful completion of this research project. Last but not the least, I would like to put a special word of appreciation, love and affection to my grand children—Eram, Zaheen, Uzair, Suhema and Affan who have been a source of encouragement and inspiration. I would like to add a special word of appreciation to beloved Affan alias Chuggarh of Imran who use to cheer, inspire, and refresh me whenever I felt tired and trophic level of mind gets jammed in writing this book.

Noor Mohammad

Contents

List of Tables

List of Figures

List of Photographs

Chapter I

Introduction

1.1 Introduction

The heavy induction of package technology in mid-sixties in agriculture popularly known as Green Revolution including modern farm implements, chemical fertilizers, high yielding varieties of seed, improved irrigation facilities, insecticides and pesticides, has brought spectacular changes in almost every parameter of agriculture. The sum total effect of all these can be gauged from the fact that the net sown area has increased from 118 million hectares in 1950-51 to 142 (46%) million hectares in 2009-10 with net irrigated area from 21 million hectares to 63 million hectares and gross irrigated area from 23 million hectares to 88 million hectares in corresponding years. Likewise gross cropped area has also increased from 132 million hectares to 194 million hectares with an intensity of cropping from 111 per cent to 138 per cent in the corresponding years. As a result the foodgrains production has gone up to about 257 million tonnes from 51 million tonnes during 2009-2010 with a compound growth rate of 3.32 over 1980-83 terannium. Consequently, India had not only become self-sufficient in foodgrains production but it has buffer stock except for only few years when the production has declined due to unfavourable environmental conditions (Vaidyanathan, 2010, Ministry of Agriculture 2008-12).

But one should not be complacent with this surplus production of foodgrains in view of the fast growing population which is likely to be 1.4 billions and 1.7 billions in the year 2025 and 2050 respectively as against one billion in 2001. It will require 260 million tonnes of foodgrains. Besides this, about 130 million tonnes of milk, 151 million tonnes of vegetables, 86 million tonnes of fruits, 10 million tonnes of meat, 10 million tonnes of fish and so on, will also be required in order to provide a balanced diet leading to better food security which is defined as "Food security exists when all people, at all times, have physical and economic access to sufficient, safe, and nutritious food to meet their dietary needs and food preferences for an active and healthy life" (FAO, 1999).

However, it is now being realized that even this existing level of agricultural transformation from indigenous agriculture to chemical, irrigation based agriculture using unscientifically heavy doses of modern agricultural inputs had unleashed many agro-ecological problems. It has severly damaged natural resource base on which it is based. The current agricultural methods based on synthetic fertilizers and chemical pest control can reap in abundance in short run at considerable economic level but it cannot be sustained in the long run, as it is socially, economically and ecologically unviable. The high-tech agriculture may create more problems than it solves and may have more questions than answers. Actually, the intensification of farming has a great natural base degradation potential. In this perspective one can think of the level of damage to environment especially in case of desertification, environmental pollution, soil erosion, salinisation and alkalinisation, waterlogging, sinking of ground water table, water and soil contamination, and deteriorating soil fertility. If the foodgrains production has to increase from 213 million tonnes to 260 million tonnes viz. increasing about 4.5 million tonnes every year in addition to a lot of non-foodgrains production, it will further accelerate the process of environmental damage. Therefore, the farming system has to be modified and improved.

1.2 Statement of Problem

Although the green revolution has brought tremendous development in agriculture especially in case of foodgrains production which has increased more than four times since independence but it has created a number of environmental problems. Therefore, there is growing awareness about the need to substitute the current production system with a more sustainable and integrated system of agricultural production. Only ushering in a second Green Revolution, where main planks would be diversification of agriculture in favour of high value commodities can fulfil the objective. The diversification of agriculture may include animal husbandry, vegetable farming, floriculture, horticulture, agro-forestry, piggries, bee farming, sericulture, aquaculture, poultry farming and so on which will not only add to the agricultural production but will also improve the quality of food.

In a scenario of a backdrop of water scarcity, particularly for surface irrigation, a chemical based agriculture cannot be relied upon. In the larger perspective of rural development and food security, it is highly desirable that a well integrated and locality specific agricultural diversification be pursued taking into account the ecological conditions and the socio-economic set-up. Only such a programme can open up the opportunity for gainful employment in rural areas and better quality of life with food security to the rural people. Hence, there is a need to research and investigate the specific agricultural diversification so as to have a balanced food security system.

Since there has been a lot of a spatio-temporal variations in the induction and consequently achievements of green revolution and its resulting environmental problems, there is a need for in depth research in this direction at meso or even at micro level. Therefore, the state of Uttarakhand has been selected for the present study due to the following reasons:

(i) The state has heterogeneous geographical environment-physical, techno-economic and socio-cultural which provides huge potential for agricultural development;
(ii) The government of the new state has formulated and implemented various programmes and policies for agricultural development, which has brought changes especially in agricultural land use.
(iii) The state has huge potential for agricultural diversification due to diversified physiographic conditions leading to biodiversity.
(iv) The southern portion of the state has practiced almost all modern agricultural inputs and has started facing its repercautions.
(v) To the best of my knowledge no work of this nature has so far been done in the state.

1.3 Research Questions

In view of the above discussed scenario, the following research questions come to in mind which needs to be researched and investigated:

- Has chemical farming really created various environmental issues?
- Is there a need for changing from chemical farming to agricultural diversification?

- Is it feasible or viable to switch over to crop diversification?
- Will the diversification of agriculture be ecologically, economically and socially viable?
- Will crop diversification help in increasing the quality of food and consequently food security?
- Will the diversification of agriculture generate more jobs for the rural masses?
- Will the new system of farming be more productive and remunerative?

1.4 Aims and Objectives

Keeping the above questions in view, the following aim and objectives have been set for the present research:

- To find out the existing pattern of landuse with special references to agricultural landuse.
- To assess the potentiality of the region for diversification of agriculture.
- To assess the job potential of agricultural diversification.
- To workout the gross and net availability of food.
- To find out the food secure and insecure regions.
- To suggest suitable measures to improve sustainable agricultural development so as to ensure food security.

1.5 Hypotheses

The following null hypotheses have been formulated in order to accomplish the present research:

- Uttarakhand has huge potential for agricultural diversification.
- Diversification of agriculture will create more employment opportunities for the rural masses of the state.
- Agricultural diversification will not only enrich the quality of food but will also bring food security in the state.

1.6 The Study Area

The state of Uttarakhand, formerly known as Uttaranchal came in to existence on 9th November 2000 in accordance with Uttar Pradesh

Reorganization of States Act 2000 as 27th Indian state and 10th in the Himalayan region. It is an intrinsic part of the youngest and highest mountain system of the world and is sometimes referred to as Kumaon Himalaya, Central Himalaya, Gurmukh, Uttarakhand and UP Himalaya. The state is conspicuously different from plain of Uttar Pradesh with respect to topography, relief, drainage, climate, soil, vegetation, language, cultural and historical background. Sprawling over 5,34,84 square kilometres it is located between 28° 45′ 10″ and 31° 27′ 42″ North latitude and 77° 34′ 10″ and 81° 02′ 11″ East longitude. The greatest breath of the region from north to south is nearly 270 kilometres, while the maximum length from east to west is about 324 kilometres between the rivers Yamuna and Kali.

It is a land locked state being bounded in the north by an international border with Tibet, which is formed by nearly 400 kilometres long mountain chain from Humla in Nepal to Shipki La in Himachal Pradesh, while the state of Uttar Pradesh lies towards the south. Its eastern and western boundaries are formed by two important river systems viz, the Kali which forms the international border with Nepal in the east, and the Yamuna and its tributary Tons demarcate the eastern boundary with the adjoining state of Himachal Pradesh.

Administratively the state presently comprises two divisions of Kumaon and Garhwal including 13 districts, 78 tahsils, and 95 development blocks. The Kumaon division comprises of 6 districts i.e. Almora, Nainital, Pithoragarh, Champawat, Udham Singh Nagar (here after U.S. Nagar) and Bageshwar, while the Garhwal division includes seven districts i.e. Dehradun, Chamoli, Hardwar, Pauri Garhwal, Tehri Garhwal, and Rudraprayag. The Garhwal division covering an area of 32,449 square kilometre (60.70% area of the state) supports a population of 49.24 lakhs (58% of the state), whereas Kumaon division spreading over an area of 21,035 square kilometre. (39.03% state area) inhabits a population of 35.65 lakhs (42.0% of state population). The state, being substantially located in the realm of the Himalayan mountain has a distinct geographical entity. It possesses the characteristics of north-south Himalayan sequence of the high and perpetually snow bound zone, flanked by the intermediate populated belt to the south and finally the lower most outlying ranges of Siwalik, which adjoins the Ganga plain. The geological structure of Uttarakhand is the outcome of three distinct and widely phases of upliftment i.e. the post-Eocene, Eocene and late Tertiary period which have resulted into four fold north-south tectonic subdivisions viz. the Trans-Himalaya, the Greater Himalaya, the Lesser Himalaya and the

Siwaliks.There is considerable variations in relief, and the altitude varies from the level plain in the south to over 7500 m along the highest peaks of the Himalayas resulting into wide range of climatic conditions. However, Uttarakhand may be divided into following 5 physiographic divisions (Valdiya, 2001 and Pant, 2001):

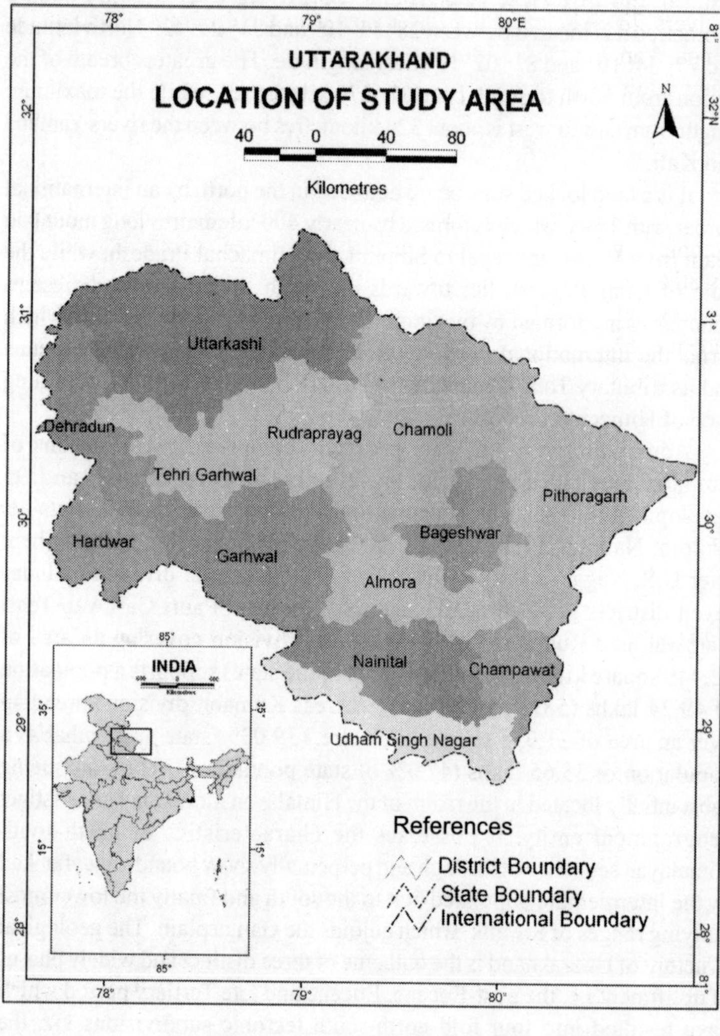

Fig. 1.1: Location of Study Area

- Trans-Himalaya
- Greater Himalaya
- Lesser Himalaya
- Siwalik Range
- Foothills and Tarai

The high Himalayan zone has extensive glaciers and snow fields like the Gangotri, the Pindari, etc. It is also source region of a large number of rivers flowing across the lower ranges in the different direction.The Ganga, Yamuna, Kali, Tons etc are the important river system which, being originated from glaciers and snow fields, are perennial rivers and provide water even during dry season due to the snow melt in the summer to the adjoining plain of the northern India. In the central region of the state there are widely distributed stretches of wide and fertile valleys originating due to river action or tectonics activities. The valleys of tectonic origin are mostly in the south lying between central and outlying Siwalik ranges. There are canoe shaped longitudinal valleys. Dehradun, Patlidun, and Kotadun are the best examples of it. The southeastern part known as *bhanbar* and *tarai* is post-Tertiary depositional phenomena. It is formed along the foot hills by the south flowing streams across the Siwalik hills. A huge amount of detritus material carried by this flow is deposited at the foot hills due to sudden drop in the gradient. The heavier and the coarse materials deposited immediately at the foot hills constitute the *bhanbar*, whereas the finer materials deposited a little farther form the *tarai*. The former is a zone of superficial absence of water, while the later is distinguished by super abundance of water.

The overall general slope in the state is from northeast to southwest with the exception of eastern portion where it is from northeast to southwest. The drainage system of the state follows the slope gradient. Majority of the rivers having their source in the snowy caps of the Himalayas are perennial in nature. There are four major river systems viz. the Ganga, Yamuna, Ramaganga (W) and Kali. Besides these, there are a number of smaller rivers, which originate in Siwaliks and are non-perennial. Ganga with its two headwaters—Bhagirathi and Alaknanda along with their tributaries is the major river system draining nearly half of the state. Kali river system with Dhauliganga, Goriganga, Ramaganga (E), and Sarju, etc. is the second important system draining nearly one-third eastern portion of the state. Yamuna with its tributaries drains a small northwestern part of the state. Ramaganga (W) originating in lesser

Himalaya flows from north to south in the southern part of Uttarakhand. However, the state has a huge potential of water resources.

The state having great physical diversities has diversified and complex climate. The undulating topography, steep slopes, deep gorges and river valleys result into great control of climate within relatively short distances. The whole range of world's climates from sub-humid to Alpine is found in the state. It may be generalized that the low lying areas (below 600 metres) have moist, hot and sub-humid climate, while it is cool temperature around the elevation of about 2000 m. At still higher elevation (up to 3500 m.) the climate is cold temperate which finally gives way to snow-line beyond which the climate becomes sub-Alpine or Alpine.

However, on the basis of seasonal distribution of temperature, rainfall and other climatic parametres, four seasons are identified in the state i.e. (i) The cold weather season (mid-December to mid-March), (ii) The Summer Season (mid-March to mid-June), (iii) The Southwest Monsoon season (mid-June to mid-September); and (iv) The Retreating Monsoon season (mid-September to mid-December). In fact the climate of Uttarakhand in a mosaic of several microclimates, and the composition and distributional pattern of flora varies in response to microclimate, edaphic, biotic and other environmental factors. However, five broad zones of vegetation types i.e. sub-tropical forests, temperate forests, sub-Alpine forests; Alpine vegetation; and Trans-Himalaya forests are found in the state.

Uttarakhand has variety of soils which may broadly be classified into two on the basis of their place of occurrence viz. deluvial soils and alluvial soils. The former in further classified into low land ferrite soil, mid-land thin and sandy soil, and the poor stony immature and unproductive soil of upper reaches. The alluvial soils may be grouped into *bhanbar*, *Tarai* and the plain soils with increasing order of fertility. The *bhanbar* soil is poor mixed with small pebbles brought down by hill streams. The *tarai* soil is well drained and fertile. It is suitable for cultivation of rice and sugarcane. The plain soil is clayey on the surface but loamy at lower levels resting on sandy sub soils. It is the most fertile soil used for the cultivation of variety of crop with high productivity.

According to 2011 census, the state supports a population of about 10.12 million with a density of 189 persons per sq. kilometre. Out of the total population, about 51 per cent are male and 49 per cent are female, while 74 per cent are rural and 26 per cent are urban (2001 census). The decadal growth of population during 1990's was 20.40 per cent, which was 24.23 per cent in the previous decade. In the first decade of 21st century it has come down to 19.20. The growth and distributions of population are highly uneven because of the highly varied physical conditions. The growth was as low as 3.67 per cent in Almora, while it was 32.88 per cent in Nainital in nineties. During 2001-2011 the growth rate of population has declined and it ranges from as low as—1.73 in Almora to as high as 32.48 in Dehradun. Similarly, the density of population ranges from 37 persons in Uttarkashi to 613 persons per sq. km. in Hardwar, while the corresponding figures in 2011 were 41 and 817 for the same districts. Only 13.4 per cent of its population lives in nearly 43 per cent of its area covered by three northern most districts, while nearly 32 per cent population lives in two southern districts of Hardwar and U.S. Nagar which cover only 9.3 per cent of its total area. The sex ratio is 963, which is below the national average. The average literacy rate is nearly 80 per cent, while nearly 88 per cent males and 71 per cent females are literate. The state is inhabited by a number of scheduled tribes, which have still retained some of their basic cultural values that have been part and parcel of their physio-cultural environment. The major tribes are Bhotia, Jaunsari, Boksha, Tharu, and Ragi. These tribes constitute about 2.5 per cent of the total population.

Uttarakhand is bestowed with diversified physical, agro-climatic, techno-economic and socio-cultural milieu. Like most other hill economies, the people of Uttarakhand also practice integrated systems of farming, forestry, horticulture, livestock and off- farm activities (Sharma and Singh, 2007). The recorded forest area constitutes about 64 per cent of the total reported area, though the actual cover based on remote sensing and satellite imageries information is only 44 per cent. The net shown area in the region is little over 13 per cent of the total reported area, although there are wide variations in it from district to district. About 33 per cent of the total area of the state is either rocky, snow covered, glaciated or otherwise unproductive or degraded land. The ecological system of the Uttarakhand is fragile and highly sensitive to any development activity that is not designed in consideration to ecology and social fabric (Bisht, 2001).

The state has basically agrarian economy engaging nearly 80 per cent of the working age-group of population. Substantial portion of the state is covered with forest followed by net sown area, which varies from as low as 3.8 per cent in Uttarkashi to as high as 53.9 per cent in Udham Singh Nagar (U.S. Nagar). Three types of cultivation are practiced here depending upon the physiographic conditions. Upraon or dry cultivation is practiced in the crest and mid-slope area with poor quality land. Almost all food grains like wheat, barley, paddy, pulses and coarse grains are grown here on terraces made across the slopes. Talaon or irrigated cultivation is practiced in the low-lying valleys of the mountain areas with most fertile deep alluvial soils. There are two main agricultural seasons, viz., kharif and rabi. Varieties of crops are grown in both the agricultural seasons. The main crops of the kharif season are paddy, pulses, chilies, fodder, etc, while wheat, barley and mustard are main rabi crops. Katil or intermittent cultivation is practiced on unterraced slopes, which have poor soils and are unirrigated. The agriculture is slowly getting diversified. Horticulture, sericulture, mushroom growing, bee keeping, dairy farming, pisciculture and herbal farming have better prospects of development in the state. (Bandooni, 2004)

The state is rich in mineral resources. Magnetite, gypsum, limestone, dolomite, shale, marbles are some of the minerals found at different location of the state. Similar is the case with industrial development. However, the state has huge potential for the development of tourism.

1.7 Review of Literature

An overview of the existing literature shows that a lot of work has been done on food security at local, national and international levels but very little work has been done in depth on crop diversification and diversification of agricultural activities with special reference to India or any of its state. Very few studies have been conducted to correlate the agricultural diversification and food security. Therefore, some of important works have been reviewed here according to their themes.

1.7.1 Conceptual Literature

The concept of food security is very complex, multi-dimensional and complicated and is debated since last three decades. Food security is generally understood to apply the arrangements whereby the people would be assured of a minimum level of foodgrains supply in a year of normal as well as poor harvest (Rentinger, 1977). Subsequently, it was realized that physical availability alone would not ensure economic access to food for all population but satisfactory production levels and stability of supplies should be matched by reduction in poverty and an increase in the effective demand to ensure economic and physical access for the poor (FAO, 1984). As such the concept of food security at the individual and household levels was taken up. Hence, according to FAO (1987) the basic concept of food security implied that "all people at all times have both physical and economic access to the basic food they need" but the World Bank (1986) has modified it indicating that "food security is access by all people at all times to enough food for an active and healthy life". However, the most comprehensive and, perhapś, largely acceptable definition came out of the World Food Summit at Rome in 1996 which states "Food security exists when all people, at all times, have physical and economic access to sufficient, safe and nutritious food to meet their dietary needs and food preferences for an active and healthy life". This definition emphasizes three critical dimensions of food security; (i) physical supply of desired food in sufficient quantity; (ii) an economic access indicating close link of purchasing power to food security; and (iii) stability in supply which includes access to global food market. A more comprehensive explanation of food security includes livelihood security for each household ensuring both physical and economic access to balanced diet, safe drinking water, environmental sanitation, primary education and basic health care (George, 1999, and Frankenberger and McCarton, 1984).

Like the concept, the indicators of food security are also complicated. Barun (1992) has suggested demand and supply of food at country level, direct comprehensive surveys of deity intake at household level and anthrometric information at individual level to measure the food security. In some studies poverty has been considered as main determinant of food security as poor do not have adequate means and entitlement to secure their access to food even if it is available in local and regional markets.

Upreti (2004) in his book "*Uttaranchal: Issues and Challenges of Development and Changes*" has critically examined various problems and challenges, which came in the way of development and change in the Uttaranchal. He has suggested pragmatic remedial measures for it.

Ruhella (2008) in his book on "*Agriculture and Food Security*" has examined the role of agriculture in food security and has concluded that agriculture plays a dominant role in maintaining the food security

1.7.2 Methodological Studies

There are very few methodological studies. Aykroyd, *et al.* (1965) worked out the co-efficient of consumption unit, which has been modified by Singh (1984). Chakravarty (1970) has worked out the co-efficient of production, which is quite prevalent.

Bora and Jalal (1995) in their article have aimed at calculating the calorie intake in relation to their main sources derived in various physical as well as the socio-economic conditions in the upper Khichari catchments area in Nainital district of Kumaon Himalayas. The methodology used by them is stratified random sampling in which they have conducted household survey and have prepared food balance sheets. The data have been converted into calories and compared with the standard requirement that is, 2430 per capita per day. It is evident from their study that the area is facing food problems both in quantity and quality as a result of which the health of the people is poor and they suffer from many nutritional deficiencies. The suggestion given by them to overcome the problem is that there should be diversification of the economy of the area by introducing horticulture, agriculture, sericulture, poultry farming, market gardening, etc.

Baba, Baba and Sanyal (2009) in their book have analysed critically the concept and status of food security, poverty level and nutritional policies in global prospectives.

Mohammad (2003) in his article "Regional Patterns of Food Security in India" has evolved a direct method of working out food security in calorific and monetary values.

1.7.3 Food Security

Hanumantha Rao and Krishnan (1997) in their article "National Food Security: A Policy Perspective for India", have discussed the policy measures to be adopted in India in order to have the national level food security. Anderson and Lorch (1999) in their article "Food Security: A Global Perspective" have tried to outline the status of the food security in a global scenario and have identified the different regions of status of food security. They have highlighted the regional variations in food security at global level and its determinants.

Sud in his study on "Population and Food Security" published in *Yojana* (2000) has highlighted that the role of the state in foodgrains management has ultimately diminished. This will have impact on foodgrain production, distribution as well as on prices with significant implications for the overall food security, especially for the poor. He has expressed his opinion that the annual compound growth rate in the post-green revolution period has been far higher for non-food crops (3.36%) than for foodgrains (2.66%). All the allied sectors of agriculture, including horticulture, animal husbandry, poultry and fisheries, have outpaced crop farming in terms of growth, but there should be no room for complacency and all attempts should be made not only to maintain the present growth but also to increase it further.

Murlidharan (2001) in his article "upholding the Right to Food" has strongly advocated that the people of the developing countries have also the right to get adequate and nutritive food. Singh (2001) in his article entitled "Human Development and Food Security" has tried to compare the human development with that of the food security in different regions. He has found a strong relationship between the food security and human development.

Swaminathan (2001)-the father of Green Revolution in India has prepared a remarkable atlas entitled *Food Insecurity Atlas of Rural India* in collaboration with the World Food Programme. He has presented it in a national consultation with the theme "Towards Hunger Free India" in which it has been visualized to eliminate poverty induced chronic hunger at the household level by the end of the Tenth Five Year Plan. The author highlights the food security as the story of poverty, discrimination and neglect of livelihoods, unhygienic living, lack of literacy, basic amenities and health care, which is the indicator of failed global and national governance.

Bhattacharya in his article "Food Security, NAP and WTO published" in *Yojana*, January (2001) has expressed his opinion that India is going to be a net importer of foodgrains in near future. Keeping this in view, developing countries like India should think of creation of a "Food Security Box" in the agreement (WTO), which may be exempted on agriculture. It would result in increased access for agricultural exports from developing countries to the markets of the developed countries. He is of the opinion that in order to have food security special attention will have to be given to the development of new crop varieties, particularly of food crops with higher nutritional values through adoption of biotechnology particularly, genetic modification, while addressing bio-safety concerns.

Mukherjee (2002) in his article has dealt with the issue of trade in food and its impact on food security. He puts forth that for India trade as a vehicle for providing food to millions and millions is not an option. The question of food security and hunger have to be viewed as the major livelihood issues confronting the nation. It must, therefore, be viewed in the larger context of international politics, sociological development, cultural needs, technological capability and sustainability of millions of peasants who have preserved the technology of growing food through low external input agriculture. Self-sufficiency in production of food and its distribution, capacity and capability of people to access food, access to basic education and primary health care for governance and free press all have strong bearing on food security and hunger.

Singh (2004) in his two volumes edited book on "*Resource Conservation and Food Security*" has included several articles dealing with food security, nutrition and health. Basu, Francis and Datta (2005) in their book have made an extensive study of the status of agriculture, food security, nutrition and health in northeastern India and have come to the conclusion that a lot has to be done in order to achieve the food security in the region. Dorin and Landy (2009) have edited a book on agriculture and food security in India in which they have included several articles dealing with status of agriculture and food security in our country.

1.7.4 Agricultural Diversification

Chaturvedi (1997) in his article has stated that Indian agriculture is once again at the crossroads. Twenty-five years after the seeds of green revolution were sown, Indian agriculture is faced with a crisis of sustainability. It is the sustainability of the production momentum achieved immediately after the green revolution had set in, that is emerging as the biggest issue confronting Indian agriculture. He states that to overcome it, there is a need for paradigm shift. The days of green revolution as the only savior for the hungry millions are over. From the time of a top-heavy model of agricultural growth, wherein technology along with essential inputs, were disseminated among farmers in the better endowed regions of the country, the 21st century will see a marked shift in agricultural growth—a growth that should possibly come from low-input agriculture based on location specific needs and appropriate technology drawing inspiration from the traditional farming practices. He further states that ensuring food security would involve adequate food production, storage, distribution and above all, capacity, particularly of the poor, in terms of money incomes to buy.

Kaur (1991) in her book *"Changing Patterns of Agricultural Land use in Bist-Doab-Punjab"* has discovered that the dynamism of agriculture is reflected not only in the changes in irrigation and land use but also in the transformation of its cropping pattern. The cropping pattern in the region has moved towards commercialization through greater stress on more paying and high yielding crops of wheat and rice. This has made the crop structure of the region more soil exhaustive. She has recommended that the future of the region lies in the diversification of agriculture and in the integrated agro-industrial development.

Shiva and Bedi (2002) in their article have examined the globalization of agriculture and its impact on food security and sustainability, and have come out to the conclusion that the green revolution model, and now the globalization of agriculture are based on the destruction of diversity, uprooting of small peasants through indebtedness, chemical intensification, capital intensification, centralization and long distance transport and thereby reducing the food security or in short it is based on the intensification of external inputs, and integration with global markets. They further state that the alternative to both the green revolution and globalization is the biodiversity intensification farming, which can only be done on small farms. Biodiversity intensification leads to internal intensification and internal integration of farms with diverse local markets. This is the real liberalization of agriculture since it frees the earth from the violence of agro-chemicals, it frees diverse species from the assault of poisons and toxics, it frees the farmer of bondage through debt and royalties, it frees the Indian consumer from having to compete with global agribusiness to access and entitlement to food.

Sundaram (2001) has presented a paper entitled "Integrating the Ecosystem Approach into Village and farm development", in the international conference held at Delhi in which he has described and explained the model of integrating ecosystem approach at village and farm level development. He has given a pragmatic suggestion for the overall sustainable development of rural India so as the people may have sustainable life style.

Alam (2003) in his M.Phil. dissertation entitled "Sustainable Agricultural Development in Nainital District" has discussed the sustainability index at block level and has identified and discussed the sustainable and unsustainable regions of the district on the basis of calorific and monetary values.

Jana (2006) in his book " Diversification in Agriculture", discussing its advantages has analysed that diversification is a paradigm shift from existing cropping pattern towards more balanced cropping system to meet ever increasing demand for cereals, oilseed, pulses, fibers, fodder, horticulture crops, etc. With globalisation of market agricultural diversification means to increase the total production in terms of quality, quantity and monetary value under different agro-climatic conditions.

Pokharel (2006) in his article "Agricultural Impressive Diversification" examines the trend of agricultural production diversification towards market oriented production of high value commercial products as against traditional consumption oriented products of low value crops. In order to analyse the trend of diversification towards high value products four key indicators have been taken viz. (i) sub-sector diversification, (ii) product diversification, (iii) agricultural export growth, and (iv) growth in agro-industries.

1.7.5 Case Studies

The studies dealing with land use and food security are varied and many. Pant, Joshi and Jalal (1991) in their article have examined the nutritional status and related aspects of agricultural landuse, land man ratio and deficiency diseases in Kotadun lying in the southeastern part of Kumaon Himalayas. The land of dun floor is intensively devoted to grain and commercial crops, while that of midslope is devoted to essential grain crops. The environmental conditions of the upland are not suitable to raise foodgrains except corn in kharif and wheat in rabi seasons. Potato and Peas are grown as commercial crops in kharif season. Food nutrients and nutritional status of the rural people have been estimated on the ICMR (Indian Council of Medical Research) guidelines. Food is usually taken without knowing its utility and nutritional importance. Generally, people are ignorant of deficiency diseases and they do not consider these as diseases. Owing to poor economic conditions and ignorance of nutritional importance, it is very difficult for a large number of people to afford protective diet and to make up the nutritional deficiency.

Chand (1977) in his book has tried to correlate the *Agricultural Diversification in the Process of Development in the Mountain Region with Special Reference to Himachal Pradesh.* In 1999 he has further extended his study at national level and has examined the potential and prospectus of a agricultural diversification and its role in developmental process (Chand, 1999).

Sharma (2002) in his edited book "Agricultural Diversification: Problem's and Prospects" has included several articles dealing with various problems and prospects of agricultural diversification in different regions. The article of Pathak (2007) evaluates the prospect of agricultural diversification in different eco-systems.

Bhole and Adhikari (1998) in their article "Rural Livelihood at Risk and How Nepalese Farmers Cope with Food Insecurity" have made an original contribution in this field by conducting a national level study for the Nepal. They have identified the food surplus and deficit regions of Nepal. The study is supported by the micro level study conducted in six villages of Nepal during 1996-97. At the end of the study they have given an Access Model summarizing the whole study and giving the future direction.

Carrera (2002) in his article entitled "Globalization, Peasant Agriculture and Food Security in Mexico: Neither Comparative Advantage nor Food Self-sufficiency" has found out that in Mexico food crisis is characterized by an increasing national supply deficit, unprecedented increases in international prices, a tripling in the value of grain and oilseed imports, a disproportionate and uncontrolled increase in consumer prices. According to him in order to overcome such food crisis there is a need for reconstruction of National Agro-food Policy to revalue and promote peasant production and survival strategies in a creative and realistic manner along with new modern and global forms of production and marketing. Pattanayak and Nayak (2005) have done the spatio-temporal analysis of crop diversification in the state of Orissa.

Joshi and Gulati (2007) have assessed the relationship between the "Agricultural Diversification and Small Holders in South Asia" and have concluded that there is positive relationship between the two. In the same book Joshi and Gulati (2007) in their article assessed the association between agricultural diversification and food security in south Asia.

Srivastava and Singh (2008) have traced the development of dairy farming in the new millennium and have found tremendous growth in it which will enhance the status of food security.

Sharma (1991) has examined the landuse in Tarai tract of Eastern Uttar Pradesh and has found out that there is considerable misuse of land in the area. There are areas where the land is either wrongly used or is not contributing as fully as it ought to have to the well-being of the people of the region. He has recommended that the land resources should be optimally and wisely used according to their capability keeping in view the local and national demands and their conservation for the generations to come. Mohammad and Mamta, 2011 has discussed the role of urban agriculture in food security in northwestern Delhi.

Pattanayak and Nayak (2005) in their article on "Crop Diversification in Orissa: A Spatio-Temporal Analysis" have studied crop concentration and diversification using Location-Quitent and have found increasing trend in crop diversification.

Although a large body of research literature dealing with food security and also few studies on agricultural diversification are available but there are very few studies which examine the co-relationship between the two. The household food security has also been studied by quite a few. There is also lack of integrated analysis that includes macro to micro level. In the present study, therefore, an attempt has been made to fill up these gaps in the field of agricultural geography.

References

Alam, S.Z., 2003: "Sustainable Agricultural Development in Nainital District", (unpublished M.Phil. Dissertation, Delhi University, Delhi).

Anderson, P.P. and Lorch, R.P., 1999: 'Food Security: A Global Perspective, in Peters, G.S. (ed.) *'Food Security, Diversification and Resource Management: Refocussing the Role of Agriculture,* Ashgate U.S.A.

Asian Development Bank, 2004: *Food Security, Poverty, and Nutrition Policy Analysis,* Oxford, New Delhi

Aykroyd, W.R, (*et al.,*),1962: *The Nutritive Value of Indian Food and Planning of Satisfactory Diet,* New Delhi.

Baba, S., Baba, S.C. and Sanyal, P., 2009: *Food Security, Poverty and Nutrition Policy Analysis,* Academic Press Pub, Burlington, U.S.A.

Barun, J.V and Bonillo, E.D. (Eds), 2009: *Globalization of Food and Agriculture and the poor,* Oxford University Press, London.

Barun, J., *et al.*, 1992: *Improving Food Security of the Poor: Concept, Policy, and Programme*, International Food Policy Research Institute, Washington

Basu, D., Kulkarni, B.F and Ray, B.D. (Eds), 2006: *Agriculture, Food Security, Nutrition and Health in North-East India*, Mittal Pub. New Delhi.

Bhattacharya, B., 2001: Food Security NAP and NATO, *Yojana,* Vol. 45, No.1.

Bisht, D.S., 2006: *Poverty, Planning and Development: A Case Study of Uttaranchal State;* Central Himalayan Institute, Dehradun.

Bisht, D.S., 2001: *Guide to Garhwal and Kumaon Hills., Uttarakhand,* Trishul Pub., Dehradun.

Bora, N.S. and Jalal, D.S., 1995: Status and Sources of Nutrition in the Kumaon Himalayas—A Case Study, *Geographical Review of India,* Vol. 57, No. I.

Chakravarty, A.K., 1970: Food Sufficiency Patterns in India; *Geographical Review,* Vol. LX, New York.

Chand, R., 1999: *Agricultural Diversification in India: Potential and Prospects*; Mittal Pub. New Delhi.

Chand, R., 1997: *Agricultural Diversification and Development of Mountain Region with Special Reference to Himachal Pradesh*, M.D. Pub. Pvt. Ltd, New Delhi.

Chaturvedi, P. (Ed.) 1997: *Reconstructing Indian Agriculture, Food Security and Panchayati Raj*, Concept Publishing Co. Pvt. Ltd., New Delhi.

Dorin, B. and Landy, (Eds), 2009: *Agricultural and Food in India*, Manohar Pub. and Distributor, New Delhi.

FAO, 1987: *The Status of Food Security in the World*, Rome.

FAO, 1999,: *The Status of Food Security in the World*, Rome.

Frankenberger, T.R. and McCarton, 1998: The Household Livelihood Concept, Food, Nutrition and Agriculture, FAO, 27

George, P.S., 1999: Some Reflections on Food Security in India, *Indian Journal of Agricultural Economics,* Vol. 54, and No.(2).

Gopalan, C., *et al.,* 2000: *Nutritive Value of Indian Foods,* Indian Council of Medical Research, Hyderabad.

Hanumantha Rao, C.H. and Radha Krishnan, R, 1997: 'National Food Security: A Policy Perspective for India' in Peters, G.S. (ed.) *'Food Security, Diversification and Resource Management: Refocusing the Role of Agriculture,* Ashgate U.S.A.

International Food Policy Research, Washington.

Jana, B.L., 2006: *Diversification in Agriculture*, Saujanya Books, Delhi.

Joshi, P.K. and Gulati, A. (Eds), 2007: Agricultural Diversification in South Asia, in Joshi, P.K., and Gulati, A.(Eds), *Agricultural Diversification and Smallholders in South Asia,* Academic Foundation, New Delhi.

Joshi, P.K., and Gulati, A. (Eds), 2007: *Agricultural Diversification and Small holders in South Asia,* Academic Foundation, New Delhi.

Masood, S. (Eds), 2009: *Peri-Urban Agriculture- Agriculture by Twenty 20 and Beyond,* APH Pub. Corp, New Delhi.

Mohammad, N. 2003: "Regional Patterns of Food Security in India" in T.P. Singh, *Resource Conservation and Food Security*, Concept Pub. Co, New Delhi.

Mohammad, N., 2002: Regional Patterns of Food Security in Rural India, *Annals of the National Association of Geographers,* India, and New Delhi.

Mohammad, N. and Mamta, A. 2011: Role of Urban Agriculture in Food Security in Northwestern India, in Singh, K.N and Rana, N.K. (Eds) *Holistic Development: A Geographical Perspectives,* R.K Book, New Delhi.

Mukherjee, A. (Ed.), 2002: International Trade and Food Security in India, *Sustainable Agriculture and Food Security: The Impact of Globalisation*, Sage Publications, New Delhi.

Murlidharan, S., 2001: Upholding the Right to Food, *Frontline,* Vol. 18 (188), New Delhi.

Pant, B.R., Joshi, R.C. and Jalal, D.S, 1991: Agricultural landuse and Nutrition in Kotadun, Kumaon Himalayas, *Geographical Review of India,* Vol. 53, and No.1.

Pant, B.R., 2000: "Geographical Personality of Uttarakhand", in Sati, M.C. and Sati, S.P. (Eds) *Uttarakhand State: Dimensions of Development*. India Pub. Co. New Delhi.

Pathak, R.K., 2002: Diversification in Different Agro-Eco-System, in Sharma, A.K. (Eds), *Agricultural Diversification: Problem and Prospects*, I.K. International Publishing House Pvt. Ltd, New Delhi.

Pattanayak, M. and Nayak, B.P.: 2005: "Crop Diversification in Orissa": A Spatio-Temporal Analysis", *Eco and WPA*, Delhi.

Pokharel, C., 2006: "Agricultural Impressive Diversification" in *Agricultural Diversification in South Asia*, Kathmandu.

Rentlinger, S, 1997: *Food Insecurity: Magnitudes and Remedies*, Washington.

Ruhela. A., 2008: *Agriculture and Food Security*, Oxford Book Co. Jaipur.

Scanes, C.G. and Miranowski, J.A. (Eds), 2004: *Perspectives in World Food and Agriculture*, Blackwell Pub.Co. Iawa

Sharma, A. and Singh, R.K. (Eds), 2007:*Composite Farming Practice and Economic Development,* Abhijeet Pub, New Delhi.

Sharma, A.K. (Ed), 2002: *Agricultural Diversification Problems and Prospects*; I.K. International Publishing House Pvt. Ltd, New Delhi,

Shrivastava, M.P. and Singh, J.S., 2008: *Dairy Development in the New Millennium.* Deep and Deep. Pub., New Delhi.

Singh, S., 2001: Human Development and Food Security, *Yojana*, Vol. 45, No.8. Food Outlook No. 4.

Sud, S., 2000: Population and Food Security; *Yojana*.

Sundaram, K.V., 2001: "Integrating the Ecosystem Approach into Village and Farm Development", in International Conference on *Sustainable Development and Sustainable Life Styles* Contributory Papers, Bhovigyam Publishers, New Delhi.

Swaminathan, M.S., 2001: Food Insecurity Atlas of Rural India, *Frontline* Vol. 18, No. 11 May 26-June 8, 2001.

Upreti, B.C, 2004: *Uttaranchal: Issues and Challenges of Development Change,* Kalinga Pub. New Delhi.

Vaidyanathan, A., 2010: *Agricultural Growth in India: Role of Technology Incentives and Institution,* Oxford University Press, New Delhi

Valdiya, K.S., 2001: Physical Division of Uttarakhand, *Current Sciences*, Vol. 81 (1). World Bank 1986 Development Report.

Chapter 2

Research Design and Methodology

2.1 Introduction

The study is based on primary and secondary data. The secondary data have been collected from published and unpublished records of the government and semi-government organisations and also from the NGO's and various websites. Some of the important latest publications used for secondary data are:

- India Meteorological Department, Pune.
- Primary Census Abstract of the districts of Uttarakhand.
- General population Tables of Uttaranchal.
- Statistical Abstract of Uttaranchal/Uttarakhand.
- Indian Agricultural Statistics Vol. I & II, Ministry of Agriculture.
- Agricultural Statistics at a glance: 2001-2012
- Forest Statistics of Uttarakhand and Forest Department, Uttarakhand

2.2 Research Design and Data Base

For the collection of primary data multistage sampling technique has been adopted. Firstly, on the basis of the physical parametres—relief, climate, drainage, soil and natural vegetation, the state has been divided into physical divisions. According to the size of physical divisions 30 villages in all have been selected from different physical units on the following basis:

(i) It should be basically an inhabited village
(ii) The villages should be representative of its physical division.

(iii) Some villages should be easily assessable, while some should be inaccessible in remote mountain area.
(iv) Some villages should be close to urban centers, while some should be away from them.
(v) Major castes and religious groups should inhabit the village.
(vi) It should have agricultural and allied activities as its major economic activity.
(vii) There should be different size of holdings.
(viii) The inhabitant of the village specially the farmers should be of different socio-economic structure.
(ix) The village should have sizeable number of households.

After deciding the criteria for selection of villages, 30 tentative locations have been identified and from each location tentatively 4-5 villages have been identified with the help of general maps, available topographic sheets and Administrative Atlas of Uttarakhand (2006).The basic characteristics of the tentatively selected village have been found out from the Primary Census Abstract of each district. At this stage 2-3 villages with their locations and characteristics have been chosen. Out of these villages finally one village has been selected after verifying all the details and their characteristics with the administrative authorities and the local people during the course of field work.

Finally from each of the final selected villages 5 to 10 per cent of the households have been selected on the basis of the socio-economic structure of the household such as age and sex composition, literacy status, size of holdings, standard of living, caste and religious composition. At the end either the head of the household or the person who takes decision in family and agricultural matters have been interviewed according to the well structured questionnaire.

A comprehensive questionnaire containing several questions on different relevant themes such as personal and household characteristics of households, farming characteristics of households, pattern of agricultural landuse, agricultural input and outputs, agricultural innovations, dairy farming, vegetable farming, horticulture, floriculture, fishing, aquaculture, poultry farming, purchase and sale of agricultural commodities, etc. have been prepared. This questionnaire has been personally, along with the project fellow have been administered in the reconnaissance survey in 4 villages viz. Mukhwa, Sheela Mala, Dumka Banger and Malsi, lying in 4 different physical divisions viz. Greater Himalayas, Lesser Himalaya, Siwaliks and foothill tarai respectively for

Table 2.1: Administrative Location of Selected Villages

Sl. No.	*Name of Village*	*Block*	*Tehsil*	*Districts*	*No. of Surveyed Households*
1.	Langasu	Karnaprayag	Karnaprayag	Chamoli	17
2.	Barhgaon	Joshi Math	Joshi Math	Chamoli	19
3.	Thewa	Raipur	Dehradun	Dehradun	12
4.	Fatehpur	Vikasnagar	Vikasnagar	Dehradun	21
5.	Rasoolpur	Roorkee	Roorkee	Hardwar	23
6.	Shahpur	Bhagwanpur	Roorkee	Hardwar	24
7.	Amsaur	Dogadda	Kotdwara	Pauri Garhwal	15
8.	Guleth	Pauri Garhwal	Pauri Garhwal	Pauri Garhwal	14
9.	Syunsi	Birokhal	Thalisain	Pauri Garhwal	59
10.	Bangar	Birokhal	Thalisain	Pauri Garhwal	16
11.	Sheela Mala	Birokhal	Thalisain	Pauri Garhwal	23
12.	Bhatwari	Ukhimath	Ukhimath	Rudraprayag	16
13.	Ratura	Augustmuni	Rudraprayag	Rudraprayag	16
14.	Manjaf	Pratapnagar	Pratapnagar	Tehri Garhwal	16
15.	Maletha	Kiritinagar	Devprayag	Tehri Garhwal	16
16.	Mukhawa	Bhatwari	Bhatwari	Uttar Kashi	16
17.	Hitaru	Dunda	Dunda	Uttar Kashi	15
18.	Kaffar Khan	Almora	Almora	Almora	17
19.	Majkhali	Ranikhet	Ranikhet	Almora	16
20.	Lohar Khet	Kapkot	Kapkot	Bageshwar	15
21.	Are	Bageshwar	Bageshwar	Bageshwar	15
22.	Mudiyani	Champawat	Champawat	Champawat	18
23.	Dugri Phartiyal	Lohaghat	Champawat	Champawat	16
24.	Dumka Bangar	Haldwani	Haldwani	Nainital	14
25.	Gotia	Kotabagh	Nainital	Nainital	12
26.	Langari	Munakot	Pithoragarh	Pithoragarh	16
27.	Kalika	Dharchula	Dharchula	Pithoragarh	16
28.	Devnagar	Sitarganj	U S Nagar	U.S. Nagar	15
29.	Motipur	Bajpur	Bajpur	U.S. Nagar	15
30.	Malsi	Rudrapur	kichha	U.S. Nagar	15
Total	—	—	—	—	**538**

limited number of households. During the course of reconnaissance survey, the discussion with local influential and literate peoples and farmers, administrative officers, local authorities, and our own experience and observation in administering the questionnaire, it has been found that the questionnaire needs to be modified and separate brief

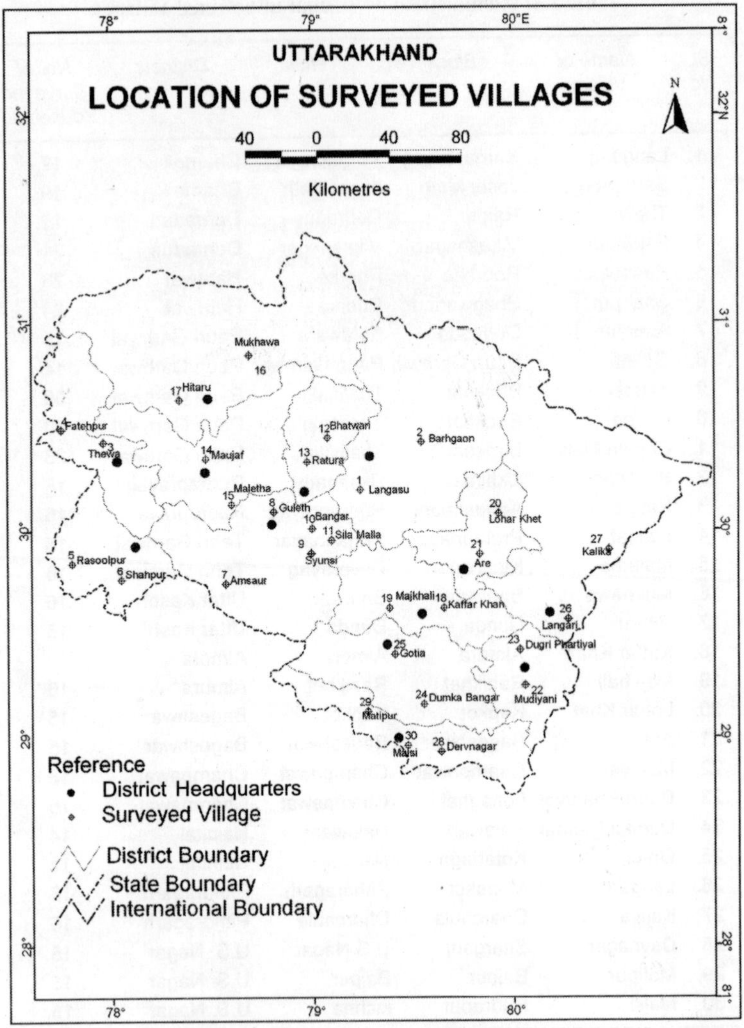

Fig. 2.1: Location of Surveyed Villages

questionnaire is required for the general information about the selected village as a whole.

Therefore, a final comprehensive questionnaire containing the following sections has been prepared (Appendix-I)

Section-I. Personal and household characteristics of households including age, sex, education and family size.

Section-II. Farming characteristics including size of holdings, patterns of agricultural land use, adoption of agricultural innovations, agricultural inputs and yield of crops.

Section-III. Agricultural diversification, which includes several question on (i) Dairy farming, (ii) Fishing, (iii) Vegetable farming, (iv) Floriculture, (v) Horticulture, (vi) Poultry farming, (vii) Aquaculture, and (viii) Apiculture.

Section-IV. Purchase of edibles by households from various sources

Section-V. Sale of edibles by households at various centres.

Section-VI. Personal experience and observation of the interviewer.

Beside this a separate questionnaire for seeking general information of the village has also been prepared (Appendix-II).

The project fellow has filled up the questionnaire in all the 30 selected villages. Many times the Principal Investigator has also accompanied the project fellow in the fieldwork.

2.3 Data Processing and its Presentation

After collecting all the relevant primary and secondary data, the processing of secondary data has been taken up first. In the first stage, the data have been stream lined and conceptual ambiguities have been removed. Most of the climatic and demographic data found in semi processed or processed form have been used with little change and calculations. The agricultural data pertaining to general landuse and agricultural landuse have been processed using simple percentage technique. Among various methods of crop combinational analysis, Doi's Critical Table for Crop-Combinational Analysis has been used in this report.(Doi's, 1959, Mohammad, 1977; Rafiullah, 1965; Weaver, 1959; Ayyer, 1969; Kostrowicki, 1972, and Majeed, 1992). The crop concentration and their diversification has been worked out using following Singh's method which is modified form of Bhatia's method.

$$C.D.I. = \frac{\sum_{i=1}^{n} xi}{n}$$

Where

C.D.I. = Crop Diversification Index

Xi = Percentage of Crops to Gross Cropped Area

n = Number of crops occupying 10 per cent or more of gross cropped area.*

* *Bhatia has considered the crops occupying more than 10 per cent of gross cropped area, while Sing has taken all crops covering more than 5 per cent of gross cropped area.*

The primary data, which are the basic foundation of this research, have been processed and analysed with utmost care and concentration. Firstly, all the entries in the questionnaire have been thoroughly checked and verified. They have been compared with the information of other villages. The entire questionnaires have been streamlined and discrepancies, if there are any, have been removed in consultation with the contact persons in the villages. If it has not been solved, the village has been revisited to verify the facts.

After streamlining the primary data, most of it has been processed using simple percentage method. The data of different nature and character have been stream lined using scoring scheme. Although, there is always an element of subjectivity in scoring scheme, but all precautions have been taken into consideration to give judiciously score to each element

The literacy level of the households and households have been ascertained using the following scoring scheme.

Literacy Standards	*No of Scores*
Illiterate	0.00
Informal education	0.25
Primary school	0.50
Middle school	0.75
High school	1.25
Senior Secondary school	1.75
Graduation	2.50
M.A and Others	5.0

Using the above scoring scheme the literacy level of each respondent has been ascertained and they have been classified in different categories. Similarly on the basis of the same scoring scheme the composite index of education of each household has been ascertained using the following formula (Mohammad, 1988 and Kanika, 2007).

$$C.L.I.H. = \frac{\sum_{i=1}^{n} x}{n}$$

Where

CLIH = Composite Literacy Index of Household

x = Total educational score of the household.

n = Total members in the households above 5 years of age.

Based on this composite literacy index of households, they have been classified in different categories.

In order to ascertain the level of innovativeness of households with regards to use of different agricultural innovations, following scoring scheme has been adopted:

Name of Innovations	*Score Value*	
	Owned	*Hired*
Tractor	3.5	1.0
Thresher	0.75	0.25
Harrow	0.75	0.25
Cultivator	0.5	0.25
Quantity of fertilizers		
< 300 kg/hec		0.25
300-600 kg/hec		0.50
>600 kg/hec		0.25
No. of each fertilizers		
No of each new variety of seeds		0.25
No of each insecticide and pesticide		0.25

After assigning the score to each respondent, his total score have been added together. Finally on this basis they have been categorized in different level of innovativeness. (Mohammad, 1976; Ramachandran, 1969 and 1992, and Verma, 2007). The status of food security has been ascertained at three levels viz.district, village and household levels. For assessing the food security at district level the study is confined to demand and supply indicators. The demand of food for the inhabitants has been ascertained by using the trennium average for the year 2003-04, 2004-05, and 2005-06 for the projected total population for the simple reason that the agricultural data at districts level is available up to 2005-06. Since, the data consists of the persons belonging to different demographic and socio-economic conditions and the requirement of food in weight or calories varies according to age, sex, occupation, body size, income, climate, etc., Singh's scale of consumption unit (0.773 or 1000 persons are equal to 773 consumption units) has been taken in this study (Singh,

1948). As such the total population of each district has been multiplied by the co-efficent of consumption unit i.e. 0.773 in order to get the consumption unit of each district.

Food availability has been ascertained by taking the edible agricultural produces at district level.The various edible imported in the district have been added, while exported items have been subtracted in order to get the gross food available. The export and import data are quite few and are not very reliable. Hence, only such data have been used which have been found reliable. However, there are many sources of leakage between the production and consumption of agricultural produce but the precise data for this are not available. Therefore, Chakravarty's (1970) formula of loss of 16.8 per cent has been accepted. Hence, his coefficient of production i.e. 0.832 has been used to calculate the net food available. In this report the net availability of food has been worked out at two levels i.e. calorific value and value added term. As all produce cannot be aggregated in terms of weight owing to their varying nature, they have been converted in to calorific value with the help of conversion table and have been added together to get the food production in calories for each district (Gopalan, 2000). The results, thus obtained, have been multiplied by the coefficient of production. i.e. 0.832 in order to get the net food available. As the nature and character of the crops grown are such that they can't be added or aggregated together, various methods such as weight, foodgrain equivalence, standard nutrition unit, calories, monetary volume, etc., have been suggested (Singh, 1972; Kendal, 1939; Khusro, 1965; Raza, 1992; Shafi, 1972; Mohammad, 2003 and Husain, 1976).The monetary value, which also has few limitations but is better than others has been taken up (Mohammad, 1992). Hence, the total agricultural produce has been converted into monetary value by multiplying the average production of crops with their state level average prices. The results thus obtained, have been multiplied by the co-efficent of production for getting net availability of food in monetary value, which ensures access to food.

Thus after ascertaining the net food available in calorific value, it has been divided by the total consumption units to get the net food available per consumption unit per annum in calorific value which has further been divided by 365 for ascertaining the availability of calories per consumption unit per day (Mohammad, 2003). The result thus obtained, has been compared with standard requirement of calories i.e. 2430 per head per day. The consumption units getting equal or more than standard requirement have been considered as food secure and the units receiving less than the standard requirements have been put under food insecure groups. However, based on it the following categories of food security/insecurity have been identified on the basis of availability of calories per day per consumption unit.

Categories	*Availability of calories*
Severely insecure	< 2129
Insecure	2130-2429
Secure	2430-3029
Moderately secure	2730-3029
Highly secure	> 3029

Similarly for value added term the assessment of food security/insecurity in monetary volume, the following categories have been identified on the basis of its availability per anum per consumption unit:

Categories	*Availability of funds in Rs.*
Severely insecure	<9000
Insecure	9000-10999
Secure	11000-12999
Moderately secure	13000-14999
Highly secure	>15000

Based on this scale the districts, village and household have been classified into different categories and reasonable accounts of each of them have been given.

The processed data have been presented in the form of tables, diagrams, maps, and charts, according to the requirement of the data using computer and GIS software.

References

Ayyar, N.P., 1969: "Crop Region of Madhya Pradesh: A Study in Methodology". *Geographical Review of India*, Vol. XXXI, No. 1, p. 17.

Bhatia, S. S., 1965: Patterns of Crop Concentration and Diversification in India, *Economic Geography'* Vol. 43.

Census of India; 2001: *Uttaranchal—Administrative*, *Atlas*, Directorate of Census.

Chakravarty, A.K., 1970: Food Sufficiency Patterns in India; *Geographical Review,* VoL LX, New York.

Doi, K., 1959: The Industrial Structure of Japanese Prefacture, *Proceeding of International Geographical Union*, Regional Conference in Japan, 1957.

Gopalan, C., *et al.*, 2000: *Nutritive Value of Indian Foods,* Indian Council of Medical Research, Hyderabad.

Husain, M., 1976: A new Approach of Measuring Agricultural Productivity in Sutlej—Ganga Plains of India, *Geographical Review of India*, Vol. 38 (3) Kolkata.

Indian Agricultural Statistics, VoL I & II, Ministry of Agriculture, Govt. of India, New Delhi.

Kendall, M.G., 1939: The Geographical Distribution of Crop Productivity in England, *Journal of Royal Statistical Society,* Vol. 162

Khusro, A.M., 1965: Measurement of Productivity at Macro and Micro Levels, *Journal of the Indian Society of Agricultural Statistics,* Vol.17 (2)

Kostrowicki, J., 1972: "Some Method and Techniques to Determine Crop and other Landuse Combinations as Used in Polish Landuse Studies", in M. Shafi, M. Anas and M.F. Siddiqui, *Symposium on Landuse in Developing Countries, 21st International Geographical Congress*, Department of Geography, A.M.U., Aligarh.

Mahmood, A., 1977: *Statistical Methods in Geographical Studies,* Rajesh Pub, New Delhi

Majeed, A., 1992: "Crop Combinational Analysis: A Review of Methodology", in Mohammad, N. (Ed) *New Dimensions in Agriculture Geography,* Volume, (5) Concepts Pub. New Delhi.

Mohammad, N., 1947: The Diffusion of an Agricultural Innovation in Haryana, *The Geographer,* VoL XXI (1), Aligarh.

Mohammad, N., 2002: Regional Patterns of Food Security in Rural India, *Annals of the National Association of Geographers,* India, New Delhi.

Mohammad, N., 2003: Regional Patterns of Food Security in India, in Singh, T.P. (Ed) *Resources Conservation and Food Security.*, Concept Publishing Company, New Delhi

Mohammad, N., 1988: *Caste and Primary Occupations: A Geographical Analysis,* Concept Pub. Co. New Delhi.

Mohammad, N., 1992: "Technological Change and Spatial Diffusion Agricultural Innovations in Trans Ghaghara Plain", in Mohammad, N. (Ed) *New Dimensions in Agriculture Geography*, Vol.(6) Concepts Pub. New Delhi.

Mohammad, N. and Raza, M., 1992: "Agricultural Meteorological Data Base and Agricultural Productivity" in Mohammad, N (Ed) *New Dimensions in Agriculture Geography*, Volume, (2), Concept Pub. New Delhi

Rafiullah, S.M., 1965: "A New Approach to Functional Classification of Towns", *The Geographer*, Vol. XII, pp. 40-53.

Ramachandran, R., 1969: *Diffusion of Agricultural Innovations in Rural India*, Mysore.

Ramachandran, R., 1992: Correlates of Innovations Among Farmers in the Ganga Plain; in Mohammad, N.(Ed) *Human Perception on Technological Change in Agriculture,* Concept Pub. Co. New Delhi

Shafi, M., 1972: Measurement of Agricultural Productivity of the Great India Plains, *The Geographer*, Vol. 19 (1).

Shah, S.L., 1985: 'Some Methodological and Pragmatic Issues in Land Capability Classification in Himalaya', in J. Singh (Ed.), *Environmental Regeneration in the Himalaya: Concept and Strategies,* Gyanodaya Prakashan, Nainital.

Singh, S,. 1948: *Concept and Computation of Consumption Units*

Singh. J., 1977: A New Technique for Measuring Agricultural Efficiency in Haryana, *The Geographer*, Vol. XIX (1), Aligarh.

Sinha, B. N, 1968: Agricultural Efficiency in India, *The Geographer,* Vol. XV, Aligarh.

Uttarakhand: Forest Statistics, Forest Department, Govt. of Uttarakhand, Dehradun.

Verma, K,. 2007: Role of Urban Agriculture in Life Support System: A Case Study of Alipur Block, Delhi. Unpublished Dissertation.Deptt. of Geography, Delhi University, Delhi.

Weaver, J.C., 1954: "Crop-Combination Regions in the Middle West", *Geographical Review,* Vol. XLIV, No. 2, p.176

Chapter 3

Geographical Profile of the Study Area

3.1 Introduction

The new state of Uttaranchal (now Uttarakhand) actually came into being on 9th November, 2000 in accordance with Uttar Pradesh Reorganization of States Act 2000 after having received the assent of the President of India, on 25th August 2000 and published in the *Gazette of India* (Extra) Part II, Section I dated 25th August, 2000 vide notification No. 37 dated August 25, 2000 (Bisht, 2001). The state has been christened as *Uttarakhand* on 1st January 2007. The state has been carved out from Uttar Pradesh by taking its hilly region including Garhwal and Kumaon divisions. In addition, the district of Hardwar has also been added in the new state. Extending over an area of 53, 484 sq km, it is situated between 28° 45′ 1″ to 31° 27′ 42″ North Lat. and 77° 34′ 10″ to 81° 92′ 11″ East Long. This land locked state is bounded in the north by international boundary with Tibet which is formed by nearly 400 km long mountain chain running from Humla in Nepal to Shipki La in Himachal Pradesh, whereas, the state of Uttar Pradesh forms its southern boundary. River Kali forms the international boundary of the state with Nepal in the east, while Yamuna and its tributary Tons demarcate its western boundary with the adjoining state of Himachal Pradesh (Pant, 2000).

Administratively, the state consists of two divisions viz. Garhwal and Kumaon including 13 districts, 78 tahsils and 95 development blocks. The Garhwal division includes 7 districts viz. Dehradun, Uttarkashi, Chamoli, Hardwar, Pauri Garhwal, Tehri Garhwal and Rudraprayag, while the Kumaon division comprises of 6 districts *viz.* Almora, Nainital, Pithoragarh, Udham Singh Nagar (hereafter U.S. Nagar), Champawat and Bageshwar. The Garhwal division covering an area of 32,449 km^2 (60.70% area of state) supports a population of 49.24 lakhs (58% of state population), while Kumaon division sprawling over an area of 21,035 sq. km. (39.30%, area of the state) inhabits a population of 35.65 lakhs (42% of the state population).

The state of Uttarakhand, lying in the Central Himalaya is the true representative of the total Himalayan geographical system. It is a panorama of an intricate mosaic of mountain ranges, hills, valleys and plains. The basic geographical framework is provided by the layout of the high Himalayan ranges which is the true orographic and hydrographic focus of the region. Thus from the highest snow bound zone in the north, the scene changes southward to numerous low elevation ridges projecting in different directions and traversed by a network of large and small perennial rivers. The Siwaliks and the narrow belt of *bhanbar*, *tarai*, and plain separate it from the alluvial plain of Uttar Pradesh (Singh, 1982).

It is a unique geographical area where altitude ranges from nearly 200 m to encapsulating series of magnificent peaks of more than 6000 m. above mean sea level. A great contrast in elevation of mountain peaks and valleys is found almost everywhere in the region. However, the state has two broad surface features i.e. the piedmont plains of bhanbar and tarai together with a small patch of upper Ganga plain in the south and the magnificent Himalayas in the north which consists of a series of parallel ranges with bold relief and complex topography. These regions vary considerably in their geographical, geo-nomic, biophysical, socio-cultural and other geo-human characteristics (Joshi, 2004; Tewari, 2006; and Pant, 2000). There are marked differences in their altitude, complex litho-tectonic feature, distinct orographic alignment, slope, structure and composition of soil, climatic condition, natural vegetation and wide range of biodiversity. These variations have performed unique function in shaping its varied natural and cultural landscape. It has a dominating role in limiting the habitation, wide range of means of livelihood, dimensions of socio-cultural and religious facts and unique spatial arrangements of the state as a whole (Valdiya, 1987 and Singh, 1982).

Geographically, the Uttarakhand Himalaya, like the rest of the Himalayas, is the outcome of the three distinct and widely separated phases of uplift from the Post-Eocene to the end of the Tertiary. The earliest phase of post-Miocene gave rise to the Greater Himalaya or the Trans-Himalaya, while the Lesser Himalaya was lifted at the end of Miocene epoch. The Main Central Thrust separates these divisions. The last phase of upliftment commenced at the end of the Tertiary period (Pleistocene) forming the Siwalik ranges in the south which are separated from Lesser Himalaya by the Main Boundary Fault (Wadia, 1960; Krishnan, 1961; Wadia and Auden, 1939; and Pascoe, 1973).

On the basis of evolutionary history, stratigraphic sequences and component rock units, geologist divide Uttarakhand into 4 traditional north-south sub-divisions viz. the Trans-Himalaya, the Greater Himalaya, the Lesser Himalaya and the Shiwalik. These geographical divisions are separated respectively from one another by Trans-Himadari Thrust (THT), Main Central Thrust (MCT), Main Boundary Thrust (MBT) and Himalayan Frontal Fault (Pant, 1995).

3.2 Physical Division

Although K.S. Valdiya (2001) has given more detailed geographical divisions based on geo-physical and cultural coherence (Table 3.1) but most commonly accepted physical divisions of the state are the following which are based on altitudinal variations, complexity of geomorphic features, geological chronology and structure of the state (Sati and Sati, 2000). These divisions are shown in Fig. 3.1.

- 3.2.1 Trans-Himalayan Zone.
- 3.2.2 Greater Himalayan Zone
- 3.2.3 Lesser Himalayan Zone
- 3.2.4 Outer Himalayan/Siwaliks Ranges.
- 3.2.5 The Foot Hills/Piedmont Belt.

3.2.1 Trans-Himalayan Zone

The Trans-Himalayan Zone is the northern most region of Uttarakhand. It is also referred to as the Tethys Himalaya or Martoli group or Indo-Tibetan Plateau or the Tibetan Watershed along the border with Tibet. It is separated by the Central crystalline by a fault-often not recognizable. The zone is made up of highly fossiliferous sedimentary formation ranging from Palaezoic to the Eocene times and is remarkably uniform in lithological sequence (Bisht, 2001 and Rawat, 2002).

This zone having an average width of nearly 35 km. is narrower in the middle and broder towards eastern and western flanks. Characterized by higher valleys, glaciers and several water bodies, it varies in elevation from 4600 to 6400 metres above mean sea level. The region is adorned with Kamet (77 56 m) the 8th highest peak of the world. It is entirely covered with deep old gorges, stones, barren rocks and snow. The major rivers like Alaknanda, Bhagirathi and Yamuna originate here. Besides these, Kali, Ramganga, Sarju, Malini, Pindar, etc. have also their source

Table 3.1: Uttarakhand: Broad Geographical Divisions

Divisions	*Altitude (m)*	*% of the total geographical area of the state (estimated)*	*Districts covered*	*Lithotectonics (Dividing surface)*
(1) Trans Himalaya	4500-5700	5.25	Pithoragarh Chamoli Uttarkashi	Trans-Himadri Fault
(2) Grater Himalaya (i) Zone of Perpetual snow (ii) Higher valleys	1800-7000 3800-7000 1800-3800	20.3 15.72 4.51	Pithoragarh Chamoli Uttarkashi Uttraprayag Tehri Bageshwar	Main Central Thrust
(3) (Middle) Lesser Himalaya (i) Higher areas (ii) Lower valleys	800-3000 1800-3000 800-1800	56.21 38.00 18.21	Pithoragarh Pauri Uttarkashi Champawat Bageshwar Almora Dehradun Tehri Nainital	Main Boundary Fault
(4) Shiwalik or outer Himalaya	300-1200	5.92	Nainital Champawat Pauri Hardwar	Himalayan Frontal Fault
(5i) Bhanbar	250-500	6.30	Nainital Pauri Hardwar	
(5ii) Tarai	200–500	4.28	U.S. Nagar Hardwar	
(5iii) Alluvial Plain	>200	1.81	Hardwar	

* Based on K.S. Valdiya's paper published in *Current Science*, Volume, 81 No. 9-10, Nov 2001.

in this region. Some south flowing rivers of Himalayas have pierced their way through the Great Himalayan ranges and it is on this basis of the river valleys alone that many groups of ranges appear distinct and also stand separated from one another. These rivers also cut these mountain ranges at certain places and form deep gorges and passes of historical importance.

As the southeast monsoon clouds cannot cross Greater Himalaya, the region lies in the rain shadow zone of the Great Himalaya. It is, therefore, a cold desert. The climate of the region is very harsh and hostile

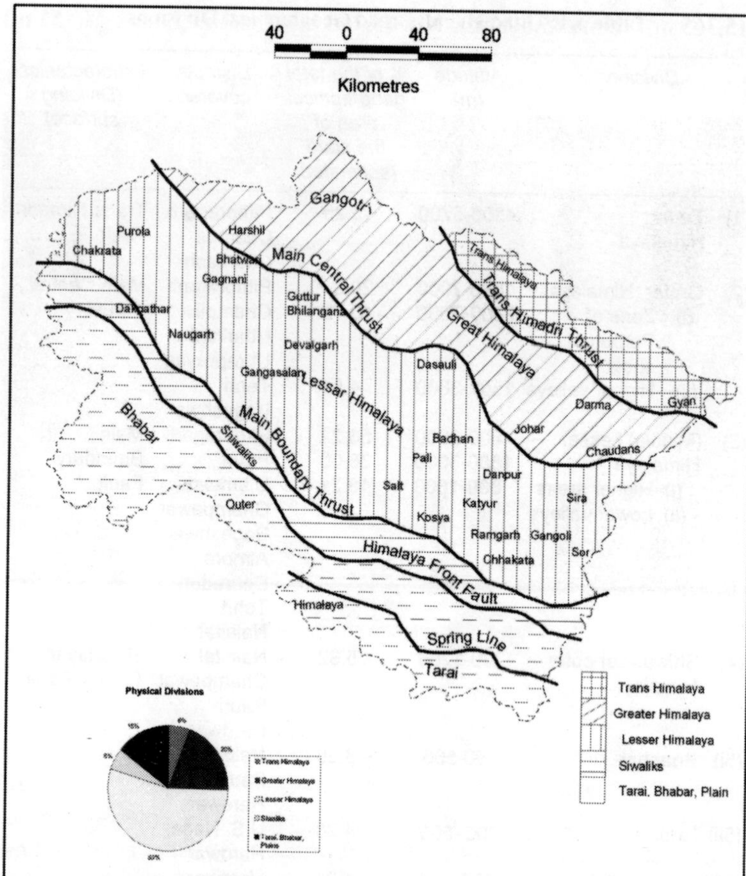

Fig. 3.1: Physical Division

with temperature generally below zero degrees celsius. It has practically little or no vegetation and habitation except 20-30 habitation of Bhotia tribes who rear their animals and feed them in the bugyals (meadows) of various sizes dotting the area. The major trade routes between India and Tibet pass through it and the Bhotia—a tribe of hard working and skillful traders, were the main settlers and traders inhabiting this tract of land. The trade links with Lhasa were mainly established in this part, and the flourishing centres for trade in wool, woolen handicrafts, semi-precious and precious stones, gold, salt, herbs, skin, animals, etc. had emerged. The main passes are Lipu Lekh (5,121 m), Bingrih (5,500 m), Barasu

(5,365 m), Mena (5,608 m), Niti (5,300 m), Lampya Dhura (5,553 m) Shalshal (7,927 m) and Darma (5,638m).

Photo 3.1: A view of Trans-Himalaya

3.2.2 Greater Himalayan Zone

The Greater Himalayan Zone also known as Himadri or Inner or Main or Higher Himalaya or Central Crystalline, is separated by a Main Central Thrust from the Lesser Himalaya. It stretches over the districts of Uttarkashi, Rudraprayag, Chamoli, Bageshwar and Pithoragarh. With a varying width of 30-60 km and elevations of 3,000 to 7,000 m, this zone is a belt of very high peaks and glacial land forms (Pandey, Swagata and Nathawat, 2011). This zone can be divided into two parts on the basis of its physiography i.e. (i) the Zone of perpetual snow, and (ii) Higher Valleys. The former is a belt of very high peaks and glacial landforms perpetually covered with snow. The prominent peaks are Nanda Devi (7,817 m)-second in India and seventh highest in the world, Kamet (7,756 m), Mana (7273), Chaukhambha Badrinath (7,138 m), Trisul (7120 m), Santopath (7075 m), Bandapoonch (6320 m), Kedarnath (6,940 m), Dunogiri (7,066 m) and Neelkanth (6,579 m), while Pindari, Gangotri, Katling and Milan are the important glaciers. Between the tree line and snow line lie the alpine meadows locally known as 'bugyals' which are the grazing lands for the herds of Bhotias, and are also an important source of rare medicinal and aromatic plants, herbs as well as the habitat of high altitude fauna. Being rich in natural resources and beauty, it holds a great promise for adventure tourism. It netsels the sacred temples and pilgrim destination of Badrinath, Kedarnath, Gangatori, Yamnotri etc.

The higher valleys lying between 1800 m to 3000 m. above mean sea level, are scattered but have permanent habitations. Bhagirathi, Maknando, Dgauli, Pindar, Gori, etc. are major river valleys from east to west. The enchanting Valley of Flowers and Nanda Devi Sanctuary are located in this region. Several Valleys of this region are inhabited by Shanka, Marcha, Talcha and Jad people of Rangpa tribes who have retained their peculiar socio-cultural characteristics. As the region is not suitable for agriculture or any permanent economic activity, inhabitant practice animal husbandry, handicraft, labour, or supplementary trade. The inhabitants are sturdy and hardworking due to harsh and inhospitable climatic conditions.

3.2.3 Lesser Himalayan Zone

The Lesser Himalaya-also called as the Himachal or the Central Zone or Lower or Middle Himalaya, extending in a varying width of 60-90 km lies between the Main Central Thrust (MCT) in the north and Main Boundary Fault/Thrust in the south with varying altitude from 800 m along lower river valleys to over 3000 m along mountain ridges close to Greater Himalaya. It consists of a series of ridges and spurs with river valleys in between and hills with varying slopes. This zone is mainly composed of crystalline and metamorphic rocks—granites, gneisses, and schsits with unfossiliferous sedimentary deposits of a very ancient age. Structurally, the rock sequence in the zone is highly disturbed being characterized by stock pilling of sedimentary and crystalline nappes. The rocks reveal closer affinity to the Pre-Palaeozoic and early Palaeozoic strata. Geologically, it is in sharp structural contrast to the Greater Himalaya in the north and Siwaliks in the south. The former reveals the existence of fossiliferous materials, while the latter shows the river borne Tertiary deposits. The intricate geology and complicated tectonics of this zone has been the subject of great controversy and has led to varying interpretations (Randhawa, 1970; Rawat, 2002 and Rawat, 2003) (Fig. 3.2).

This zone, which is a true representative of the state is not only from economic point of view but also from social and cultural perspective, can be divided into two sub divisions i.e. (i) Higher Hills, and (ii) Lower

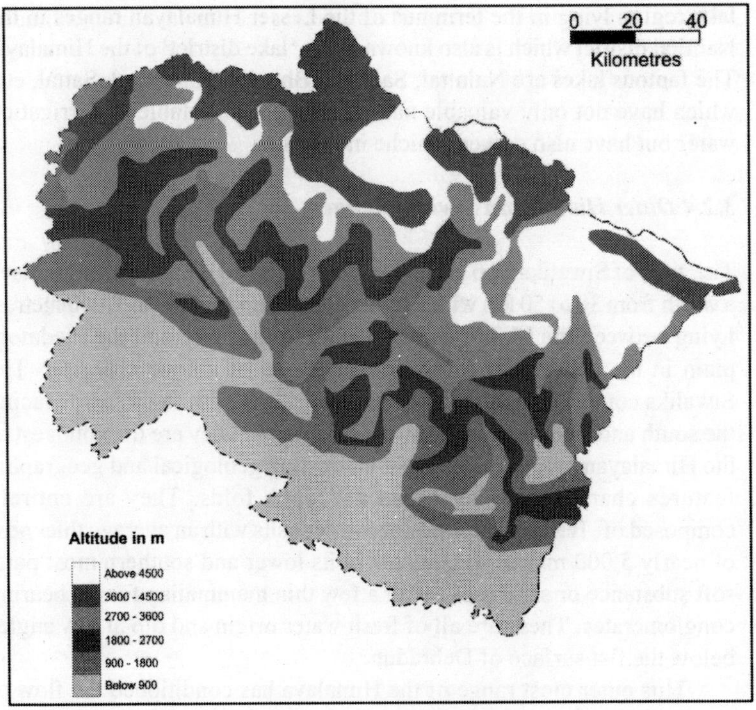

Fig. 3.2: Physiography

Valleys. The Higher Hills rising upto a height of nearly 2,500 m includes hills and mountain ranges of different sizes and are mostly covered with tropical, sub-tropical to temperate alpine forest. By virtue of its rich natural resources and scenic beauty, it holds a great promise for the development of tourism and related activities (Bhakuni, 2000). Almost all the important hill stations of the state viz. Nainital, Almora, Pithoragarh, Ranikhet, Gopeshwar, Srinagar, Mussoorie, Uttarkashi lie in this zone. The Higher Valleys (800-1,600 m) are densely inhabited areas due to their fertile and deep soils, adequate water for irrigation, warmer climate and other favourable geophysical and geo-economic conditions. It has always been an area of major interest for Kings and British. The zone has a dense network of major and small river valleys with well marked basins such as the Saryu-Kali basin, the Ramganga and Kosi Basin, the Alaknanda-Bhagirathi basin, and Tons- Yamuna basin.

The peculiar feature of this zone is the occurrence of a crescent shape lake region lying in the terminus of the Lesser Himalayan ranges in the Nainital district which is also known as the 'lake district' of the Himalaya. The famous lakes are Nainital, Sadiatal, Bhimtal, Khurpatal, Sattal, etc. which have not only valuable natural reserves of potable and irrigation water but have also carved a niche in tourism.

3.2.4 Outer Himalayan Siwalik Ranges

The zone of Siwaliks also known as Outer or Sub-Himalayan ranges have a width from 30 to 50 km with varying elevation from 200 to 1,000 metres. Lying between the Main Boundary Fault in the north and the Piedmont plain in the south, it is a transitional region of unique character. The Siwaliks comprises of a series of parallel ridges with steep scarps facing the south and a gentles slope towards the north. They are the youngest of the Himalayan ranges and present interesting geological and geographic features characterized by broad and open folds. They are entirely composed of Tertiary and Pliestocene deposits with an average thickness of nearly 5,000 metres. It contains in its lower and southern most parts soft substance or sand rocks with a few thin mammalian fossils bearing conglomerates. These are all of fresh water origin and dip at low angles below the flat surface of Dehradun.

This outer most range of the Himalaya has conditioned the flow of rivers flowing from north. Almost every river of the state flowing downward from the Lesser Himalaya follows a long course parallel to Siwalik ranges which afford a passage to the rivers only at the points of their alignment alteration. Ganga, Ramganga, Kosi, etc. are best examples. These ranges affect the climatic conditions specially the pattern of rainfall distribution along these ranges and the interior areas. The region experiences tıopical climatic conditions and is abode of several species of flora and fauna.

Although the Siwalik ranges are not contiguous to the Lesser Himalayas and are considerably apart, longitudinal valleys occur between these two physiographic regions. These canoe-shaped structural valleys are called 'The Dun" or Dwars which occur in the north of Siwaliks and do not from a continuous series. They are more conspicuous in the western part and are almost absent towards the east. The Dun of Dehra (Dehra Dun) sprawling over 25 km wide and 75 km long belt with elevation ranging from 660 m to 300 m valong Yamuna and Ganga is the biggest and most famous developed "Dun. Kotri Dun lying in the southern part of Pauri Gahwal, Patli Dun between Ramganga and Kosi, Kota Dun between Kosi and Baur, Chaukhamba Dun between Ganga and Ramganga are the other important 'Duns' of the state. Due to their typical geographical structure, these Duns have abundant flat surface, highly fertile soil together with warm tropical climate and continuous water supply which encourages intensive cultivation specially rice which is internationally famous for its quality and taste. These Duns are densily populated with large urban conglomerations.

3.2.5 Foothills or Piedmont Belt

Broadly it consists of three physiographic units viz. the bhanbar, the tarai and the plain lying between the Siwalik and the Ganga plain. In the south of Siwalik lies the 10-20 km narrow belt of Bhanbar running from east to west with a unique continuity. The Himalayan torrents rush down to this belt from the steep slopes and disappear under boulder, course sand, pebble and gravels due to its extereme porosity.

It is covered with thick tropical deciduous commercial forest. Sal, Shesham, Semal, Khair are valuable species found here. It is unsuitable for agricultural activities due to lack of recognizable soil base. It lies between 250 and 500 m above mean seal level and covers nearly 6.3 per cent of the state areas extending over Nainital, Pauri Garhwal and Hardwar.

South of the Bhanbar lies the Tarai which is characterized by a high water table often referred to as the spring line where the disappeared water in the porous strata of bhanbar reappears on the surface due to change in structural conditions and it becomes a zone of superabundance in surface water. It has a specific physical and cultural landscape due to its level plain, fertile soil with excessive moisture and rich humus together with high water table. Previously, this belt consisted of thick forests and rich fauna, and was inhospitable owing to scourge of malaria and vagaries of climate but now due to widespread drainage control and other land reclamation schemes it has turned into a zone of flourishing agriculture.

A small portion of the state lying in the southwestern margin constitutes the part of upper stretch of Ganga plain. It is difficult to draw a sharp boundary between the parallel strip of tarai and plain, but the micro physiographic features such as—cone and inter-cones, bluffs, levees, bhurs, old terraces and riverine tracts, physiographically separate the plains from adjoining tarai. It is a homogenous level plain with uniform surface materials. Lithiologically it constitutes the Khaddar-low lying area along the river course, and bhangar-upland area. Pleistocene sand ridges locally called as bhur is another distinct feature.

The *tarai-bhanbar* belt has gained a great importance today. It has been gradually reclaimed during the last 50 years for agriculture due to induction of new agricultural techniques by the enterprising hardworking refugees from Pakistan and Bangladesh, and retired army personnel's and freedom fighters of Uttarakhand. Irrigation facilities are adequate and the agricultural productivity is pretty much higher than the upper hilly region of the state.

3.3 Drainage System

The state of Uttarakhand has a dense network of perennial and seasonal rivers. However, it is mainly drained by the large number of perennial rivers originating from the snowy covers of the Greater Himalaya. A few rivers rise from the high forested zone of Lesser Himalaya, and they usually carry a small amount of water except during the raining season. In addition to these, there are small seasonal rivers and rivulets like Dhela, Bhalora, Nandhaur, Lodhiya, etc. originating along the Siwalik ranges.

However, the state is mainly drained by 4 major river systems. viz. Ganga and its tributaries; Yamuna and its tributaries; Ramganga; and Kali (Fig. 3.3).

3.3.1 Ganga River System

Ganga being the main river, is the largest drainage basin in the state in terms of tributaries and areal coverage. Ganga-most prominent and holy among all rivers, represents the soul of India, her rich culture, history and civilization. Bhagirathi and Alaknanda which originate in the glaciated zone of the Greater Himalaya are the two main headwater streams of Ganga in the northern slopes of Trishul peak. These headwater streams confluence at Devprayag from where the stream is known as Ganga all along its downstream course. The holy and legendry river Bhagirathi named after King Bhagirath has its source at Gaumukh (the shape of ice formation is like a cow's mouth) where the mighty river emerges from the depth of the Gangotri glaciers. Bhagirathi is also fed by several streams such as Jadganga Jodhganga, Rangmech, Neelapani, Ghora, etc. which have their source in the trans-Himalayan zone. The river flows along vast stretches of level land between Uttarkashi and Tehri Garhwal where the valley is open and wide and supports a flourishing agriculture.

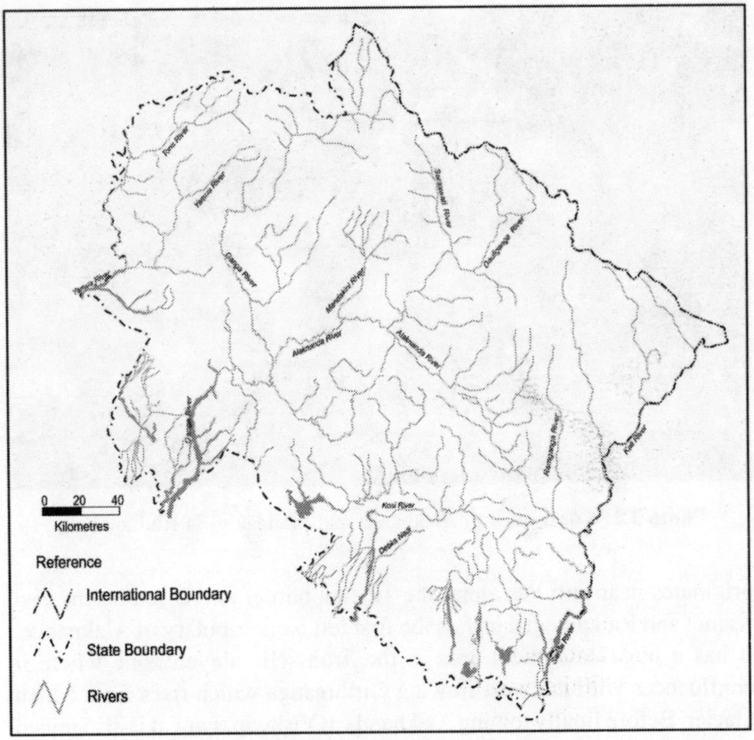

Fig. 3.3: Drainage Pattern

Bhillanganga, originating near Khatling glacier in the extreme north eastern part of the Tehri district is a major left bank tributary of Bhagirathi. It receives a combined flow of its two right-bank tributaries Dharang and Balganga near Ghansyali. The Bhillanganga having a large catchments area extending over the northern and northeastern parts of Tehri district, ultimately joins Bhagirathi near Tehri.

Alaknanda—the eastern headwater of Ganga, rises from Santopanth glacier near Badrinath and flows a distance of nearly 195 km across a highly rugged topography in the district of Chamoli, before its joins Bhagirathi at Devprayag. It has several feeding streams like Sarswati, Khirganga, Girthiganga Rishiganga, Kiogad, Ga.ıeshganga, etc. which flow from the Kamet-Mrigthuni Gauri Parbat and the valley of flowers-Kemkund- Hathi Parbat Area.

Sarswati, emanating from Alkapuri bank, is the northern most tributary of Alaknand. It joins at Mana village. Dhauliganga which

Photo 3.2: Confluence of Alaknanda and Mandakini in Rudraprayag

originates near Niti Pas along the Tibetan border and is fed by the east Kamet and Raikana glaciers, is the first left bank tributary of Alaknanda. It has a huge catchment area in the Trans-Himalayan zone where it confluences with the west flowing Girthiganga which rises from Milan glacier. Before finally joining Alaknanda at Vishnuprayag, it is also joined by Rishiganga. Nandakin, rising near Rupkhand is another left bank tributary which joins it at Nandprayag. The Pinder which originates from Pindari glacier, is a much larger south bank stream of Alaknanda. It has large number of tributaries all along its course such as Kapbni, Sundardhunga, Kailganga, Pranmuti Ata, etc. The river joins Alaknanda at Karanprayag after flowing a mountainous course of nearly 105 km. From Karanprayag Alaknanda flows westward. Mandakni, rising from the glaciated zone near Kedarnath is a major right bank tributary of Alaknanda. It has a large number feeder tributaries such as Vasukiganga, Kaliganga, Mandhani Ganga, etc. in its upper course, while Lastar Gad and Hilang nadi are major downstream tributaries. The river flowing all along southerly direction joins Alaknanda at Rudraprayag from where Alaknanda does not have any important tributary till Devprayag except very small streams. Vishnuprayag, Nandprayag, Karanprayag, Rudraprayag, and Devprayag are together called Panchprayag-confluence of five sacred rivers (five holy confluences). Covering a length of 66 km from Devprayag to Hardwar, the combined flow of two rivers-Bhagirathi and Alaknanda assume a large size and carries a significant volume of water. The major tributaries joining along this stretch are Nayor, Song, and Hyuni (Fig. 3.3).

3.3.2 Yamuna River System

River Yamuna is the major stream of this system, which sprawls over the northeastern part of the state. It originates from Yamnotri glacier and Saptrishi Kund on southwestern slope of Bandar Poonch peak. Hanumanganga and Sauri are important headwater streams in its upper course. Flowing in southwesterly direction, the river finally enters in the plain at Dhalipur after traversing a distance of 136 km in the state. River Tons originating from northeastern slope of Bandarpuch, is a major tributary of Yamuna. Rupin is its main feeder in the upper course. It has a catchments area in extreme northwest of Uttarkashi district and also forms the boundary between Himachal Pradesh and Uttarkashi for some distance. Finally it confluences with Yamuna, a little downstream of Kalsi (Fig. 3.3).

3.3.3 Ramganga (W) River System

River Ramaganga, the main stream of this system originates along the southern aspects of the Dudhatal slopes in Chamoli district. It has many seasonal tributaries such as Ganga, Sona, Devgad, Mandal, Naurar, Nayar etc. which have their sources in Lesser Himalaya. River Kosi which joins the river outside the state is also a tributary of Ramganga. It has its source region in the Lesser Himalaya along the high forested ranges of Bhatkal northwest of Kausani. Flowing in a tortuous course in the districts of Almora and Nainital the river enters the plain at Ramanagar. It is subject to violent floods causing emmence damage to life and property. It has a great erosive power and changes its course very frequently. Besides this, Gaula, Dabka are other tributaries of Kosi (Fig. 3.3).

3.3.4 Kali River System

River Kali (also known as Sharda river) is the trunk river of the system. It originates near Lepulekh Pass where the boundaries of Nepal, India and Tibet meet. Forming the International boundary of eastern Uttarakhand and western Nepal, it is the largest river of Uttarakhand running a course of about 252 km from it source to Tanakpur where it enters the plain. Dhauli (k), Goriganga, Kuthiganga which have their source in Trans-Himalaya Zone are its main feeders in its upper course. Dhauliganga (K) has two head water streams i.e. Dharamganga and Lisser which, rising in snow bound zone of Tibetan border, provide huge volume of water. Finally Dhauli (K) Joins Kali at Tawaghat. Goriganga—an important right bank tributary of Kali, originates in Milan glacier and flows between two highland zones of snow i.e. Nandakat group towards west and Punchchuli group towards east in its upper course. As such it has a considerable drainage area and water discharge. Sarju—another major right bank tributary of Kali river originates from the glaciated zone of Nanda Kat group of ranges and is one of the largest tributary traversing a course of about 146 km. from Bhadratung to Pancheshwar where it meets Kali. Sarju has vast fertile agricultural tract along its bank at several places which supports a large population. Besides these, Gomati, Ramganga (E), and Lodhiya are other notable tributaries of Kali. Several rivers originating in Nepal also join Kali on it left bank. Among them Chamoli is most prominent (Fig. 3.3).

Besides these, there are innumerable small streams, rivulets and big and small lakes spread over the entire state of Uttarakhand.

3.4 Climate

The state of Uttarakhand has a great physical diversities ranging from plains to over 7000m ranges of Greater Himalayas with undulating topography, steep slopes, deep gorges, waterfalls, escarpments and river valleys which results into complex and contrasting climates within a short distance. Similarly, the varying combinations of varied temperature, humidity and precipitation conditions create a bewildering confusing variety of climates in a single mountain range. Thus the whole range of world's climate from tropical, sub-tropical, cool temperate, cold temperate, sub alpine to Alpine are found if one moves from foothill plains to the perpetually snow bound zone of the Greater Himalayas. The mountains of various heights generally have wide difference of climatic conditions due to their elevation effect, alignment of ranges, leeward and windward locations, proximity to snow cover, water bodies and vast stretch of vegetation cover. Besides these, the local relief which effectively determines the insolation on slopes with varying exposure, aspects and variation of exposures to winds also, results into micro climatic differences over short distances. Thus, the region represents a mosaic with several micro-climates fitted into the patterns of altitude and relief. (Shah and Rawat, 2005).

However, Uttarakhand, on the whole, experiences monsoon type of climate characterized by complete reversal of the winds. Like whole of India, here also the year is divided into the following four main seasons on the basis of temperature, pressure, wind direction, humidity, rainfall and other broad weather elements.

3.4.1 Cold Weather Season.
3.4.2 Summer Season
3.4.3 Rainy Season
3.4.4 Retreating Monsoon Season.

3.4.1 Cold Weather Season

Cold weather season, also known as winter season, extends from mid-December to the middle of March and is the coldest season of the year. The mean monthly temperature varies from as low as 7° C in Lohaghat and Joshimath to as high as 21° C in Ramnagar and Kiccha in the month of January. The temperature starts gradually increasing and reaches to the lowest figure of 11°C at Lohaghat and Mukteshwar and the highest figure of 32° C at Kichha during the month of March. The relative humidity in fairly high with a lot of local variations. The wind generally blow from west to east. The region receives precipitation mainly due to western disturbances which are of sub-tropical origin. These cyclonic disturbances, originating in the Mediterranean sea, reach India passing through Turkey, Iraq, Iran, Afghanistan and Pakistan, and are the main source of precipitation in the northern plains of India and the entire Western Himalayas. During this reason, the precipitation mainly occurs in the form of snow in the higher areas of the Himalayan ranges, while light to moderate rain is received in the foothills and inner areas. However, it generally decreases from west to east.

The distributional pattern of precipitation during winter season clearly indicates that the Siwaliks and the foot hills receive nearly 100 mm of precipitation but it increases towards north and northeast. Mussoorie and Rajgarh together with their surroundings receive more than 400 mm of precipitation during this season. The substantial portion of the state specially the eastern part receives around 200 mm of precipitation. The winter rainfall varies from about 8 per cent at Nainital to 29 per cent at Joshimath from their respective mean annual precipitation

UTTARAKHAND: ENVIRONMENT AND HUMAN ACTIVITIES

Generalised View of Interrelationship

Vegetation Zone	Altitude in metres	CLIMATIC ZONES	Mean annual temperature in °C	
	5000		–0.2	
Snow line		ARCTIC OR FROZEN		
(Short and soft stemed plants	4650		0.0	Upper limit of monsoon grazing
High altitude midows (Bughiyls)	4500		0.1	
Sedge and grasses Herbs Shrubs Timber Line	4000		3.62	
	3500	GLACIAL	6.3	
Upper limit of forests Rhododendron Betula Birch				
Ablea (Sub alpine) Silver fir Blue pine, fir	3000	ALPINE	9.1	Upper limit of permanent
Cypress, Deodar Kharsu, Monu Temperate Coniferous ever green Kharsu	2500	SUBALPINE	11.8	Upper limit of successful wheat cultivation of wheat, coarse grain
Moru Banj-oak Temperate Moist broad leafed	2000	COOL TEMPRATE WARM TEMPERATE	14.5	Fatar and potato summer crop Upper limit of successful paddy cultivation Wheat paddy
Banj-oak Chir Pine (low montane needle leaf forest)	1500	SUB-TROPICAL	17.2	Manduwa, Jhingora Chuwat, Gahat, Bhat Potato Vegetables
Sal forest	1000		20.0	Upper limit of double rice crop Paddy Wheat Sugarcane Potato
Mixed hard wood (Khair, Shishum, Teak)	500	TROPICAL	23.7	Vegetable all seasons

Fig. 3.4: Environment and Economic Activity

Table 3.2: Uttarakhand: Mean Max, Mean Min. Temperatures °C and Rainfall

Month	*Tem/Rainfall*	*Dehradun*	*Mukteshwar*	*Nainital*	*Roorkee*
January	Mean.Max.Temp	19.4	10.7	10.7	20.4
	Mean, Min.Temp	5.8	1.5	1.7	6.8
	Rainfall	51.60	54.50	81.10	36.90
February	Mean.Max.Temp	21.9	12.5	12.3	23.5
	Mean, Min.Temp	7.9	2.8	3.5	8.2
	Rainfall	48.10	46.40	7.00	33.00
March	Mean.Max.Temp	26.7	16.7	16.3	29.1
	Mean, Min.Temp	12.2	6.2	7.5	12.8
	Rainfall	51.30	55.70	54.40	34.80
April	Mean.Max.Temp	32.3	21.3	20.0	35.7
	Mean, Min.Temp	16.9	10.4	11.9	18.3
	Rainfall	17.30	27.60	32.20	8.40
May	Mean.Max.Temp	35.8	23.9	23.5	39.2
	Mean, Min.Temp	20.5	12.9	14.6	22.1
	Rainfall	36.20	58.00	76.90	19.20
June	Mean.Max.Temp	34.9	23.5	23.5	38.1
	Mean, Min.Temp	23.0	14.3	16.4	24.9
	Rainfall	225.00	131.60	331.80	128.70
July	Mean.Max.Temp	30.5	20.9	26.6	33.5
	Mean, Min.Temp	22.5	14.2	16.5	24.8
	Rainfall	718.60	310.60	759.30	342.60
August	Mean.Max.Temp	29.5	20.4	21.0	32.4
	Mean Min. Temp	22.0	14.0	16.0	24.5
	Rainfall	734.00	298.80	587.10	336.80
September	Mean.Max.Temp	29.7	20.3	20.7	32.7
	Mean, Min.Temp	20.6	12.5	14.1	22.9
	Rainfall	321.70	209.70	348.90	157.80
October	Mean.Max.Temp	28.5	19.3	18.7	31.4
	Mean, Min.Temp	15.9	9.5	9.7	17.2
	Rainfall	47.50	78.70	150.70	38.40
November	Mean.Max.Temp	25.1	16.4	15.4	27.1
	Mean, Min.Temp	10.4	6.0	5.7	10.5
	Rainfall	11.30	6.30	6.90	5.50
December	Mean.Max.Temp	31.3	13.5	12.9	22.3
	Mean, Min.Temp	6.8	3.5	3.1	6.7
	Rainfall	23.20	18.60	26.90	14.30

Source: India Meteorological Tables-1951-1980, India. Meteorology Office, New Delhi,1999

3.4.2 Summer Season

The summer season—also known as pre-monsoon season begins from mid-March and continues till mid-June. The temperature rises abruptly during this season reaching a high of about 40°C at Tanakpur, Ramnagar and Kiccha during the months of May and June which are the hottest months of the year. The lowest temperature of around 19°C is recorded at Ranikhet and Lohaghat during the same months. In general the temperature is higher in foothills and southern flank of the Siwaliks and it goes on decreasing towards north in the higher Himalayas. The relative humidity is low. It is usually a dry period characterised by calm winds and sunny weather which provides an ideal condition for tourist activities and holidaying in this region. Occasional showers and hailstorms in association with thunderstorm occur during this period. The mean monthly rainfall during this season varies from nearly 11 per cent at Roorkee, Dehradun, Mussoorie, Hardwar and Tehri Garhwal to 23 per cent at Pithoragarh.

3.4.3 Southwest Monsoon Season

This is a season of general rains which begins from mid-June when the monsoon strikes the Siwalik ranges giving the first impact of monsoon rain. The temperature starts decreasing from June and the lowest temperature of nearly 18°C in recorded at Srinagar, Mussoorie and Mukteshwar. However, the average temperature during this period is nearly 25°C. The relative humidity is extremely high ranging from 80 to 90 per cent. Fig 3.5 representing the distribution of rainfall during monsoon season shows that the western plains and the foothills receive nearly 75 cm rainfall, whereas the Lesser Himalayan ranges get 100 to 175 cm of rainfall. The Mussoorie and its surrounding receive more than 700 cm rainfall, while the extreme north and northeastern part get less than 50 cm rainfall. The areas around Nandprayag and Baranag receive more than 100 cm rainfall. This is the rainiest season and nearly 70 per cent of the total rainfall is received during this period. July and August are the rainiest months with high intensity of rainfall leading to occasional cloud burst which results into flash flood causing extensive damage to life and property.

3.4.4 Retreating Monsoon Season

This season-also known as post-monsoon season begins from mid-September and continues till mid-December. The temperature further decreases reaching to as low as 11°C at many place e.g. Almora, Dharchula, Ranikhet, Pauri, etc. The relative humidity is low and complete reversal of wind direction takes place. This is a period of dry weather receiving only 3 per cent of the total rainfall. November is the driest month receiving not more than one cm of rainfall anywhere in the state.

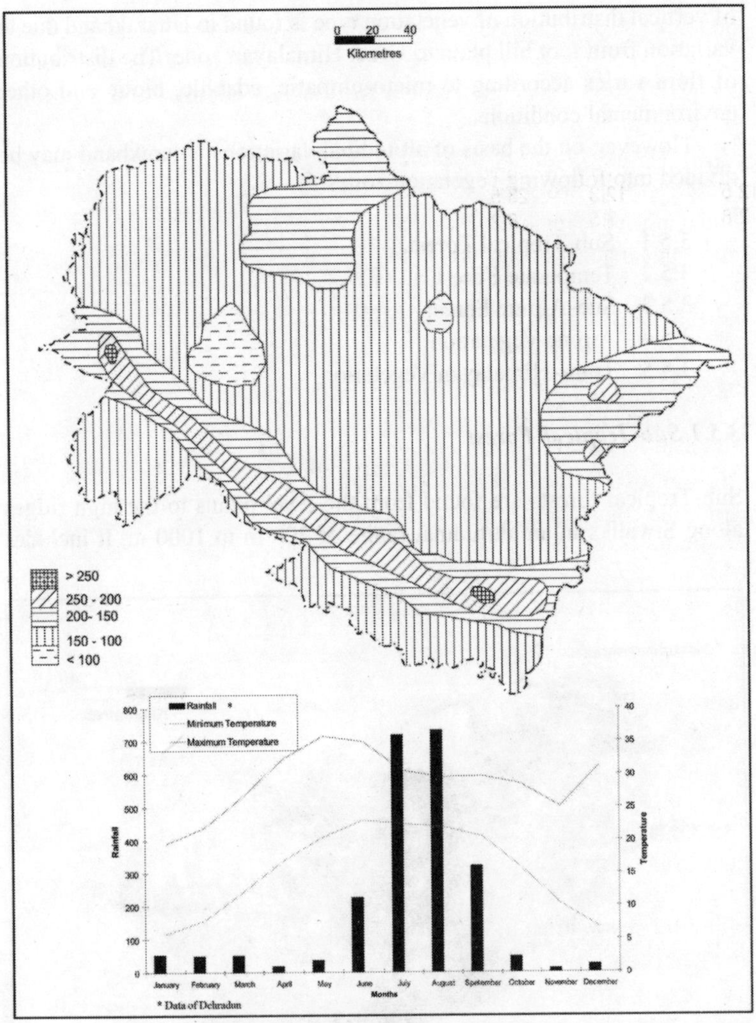

Fig. 3.5: Rainfall and Temperature

3.5 Natural Vegetation

Altitude, climate and soil are the basic determinants of the natural vegetation and all these three are highly complex and variable in the state of Uttarakhand. The altitudinal variations affect the climatic parametres which in turn influence the vegetation types. Hence, a bewildering complex of vertical distribution of vegetation type is found in Uttarakhand due to variation from foot hill plain to Great Himalayan zone. The distribution of flora varies according to micro-climatic, edaphic, biotic and other environmental conditions.

However, on the basis of altitudinal variations Uttarakhand may be divided into following vegetation zones:

- 3.5.1 Sub-Tropical Forest
- 3.5.2 Temperate Forest
- 3.5.3 Sub-Alpine Forest
- 3.5.4 Alpine Vegetation
- 3.5.5 Trans-Himalayan Vegetation

3.5.1 Sub-Tropical Forest

Sub-Tropical forests are found from the level plains to the high ridges along Siwaliks in an altitudinal range of 200 m to 1000 m. It includes

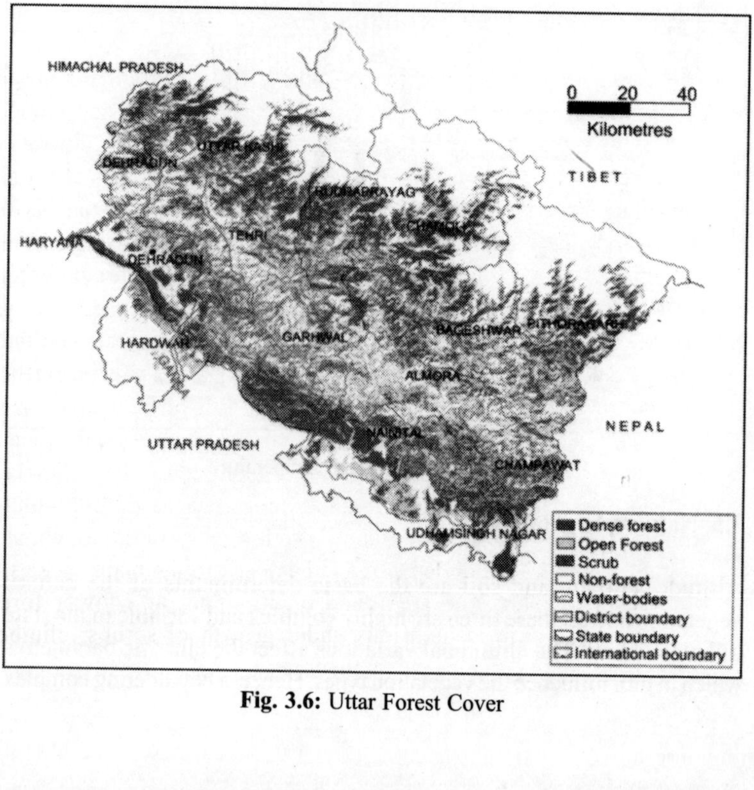

Fig. 3.6: Uttar Forest Cover

three sub-types i.e. Savana vegetation, sub-tropical deciduous forest and low altitude mountain forest. The Savanna vegetation is mainly found in the tarai of U.S. Nagar and limited stretches in the plain towards east. It mainly consists of tall grasses and mixed tree species of bamboo, dhak, jamun, kachnar, semal and peepal together with eucalyptus. In addition, poplar, sal and teak are also found but the growth of sal with respect to quality and height is poor. The sub-tropical deciduous forests are found in the bhanbar and Dun valley areas between 250-500 m elevation. They generally consists of thick patches with a dense and fast growing undergrowth of grasses, creepers and bushes. Shisham, kair, sal, bamboo, eucalyptus are the major species. But haldu, semal, dudhi, amaltas, bel, ghingan are also found all along the river valleys. The two main world life sanctuaries—the Rajaji National Park and Carbet Tiger Reserve are located within this forest zone. The low altitude mountain zone mainly occurs on the mid and lower slopes of Siwalik ranges and low lying valleys. A large variety of species of forests together with variety of miscellaneous broad leaved species with various kinds of shrubs are found here.

3.5.2 Temperate Forest

This forest zone is found between 1000-2000 m altitudinal zones and constitutes three sub-types viz. warm temperature forests, temperate coniferous forests and cold temperate forests. The warm temperate are found between 1000 and 1300 m height along higher ridges of Siwaliks and adjacent slopes of lesser Himalaya. It is a transitional zone between higher and lower zones with mixed species having hill sals and pine. Chir mixed with sal is found at higher elevations. Temperate conifer forest is an exclusive zone of chir forest sometimes interspersed with pine banjoak on drier slopes of Lesser Himalaya. The temperate coniferous forests occur over an altitudinal range of 1300–1800 m over an extensive area on the drier slopes of Lesser Himalaya. It is an exclusive zone of chir forest with practically no undergrowth but at places chir- pine are also noticed. Cold temprate forests occur along the Lesser Himalayan mid-slope and interior valleys in an altitudinal range of 1800-2000 m which is characterized by low rainfall and low temperature for about six months in a year. Rhododendron and banjoak are the main species but blue-pine and chir-pine are found on its upper and lower forests respectively. There is luxurious under growth of scrubs, climbers, herbaceous, ferns, orchids etc. in the forests.

3.5.3 Sub-Alpine Forests

Sub- Alpine forests occur between the altitudinal ranges of 3000-3500 m in the lower slopes and valleys of Greater Himalaya. There are two broad sub-types of these forests viz. sub-alpine moist deciduous and sub-alpine high altitude conifers. The farmer is found along some lower valleys and lower slopes of Greater Himalaya. Blue pine, low level silver fir, cypress moru, deodar are the principal species but great variety of broad leaved species of walnut, ash, elm and mapple are also found with numerous climbers, scrubs and herbs. Confined over an elevation between 2500 and 3500 m are sub-alpine high altitude conifers with rhododendron, blue pine high altitude silver fir, birch as major species. Besides these, oak, spruce, yeh, muple are the other species found in different altitudinal zones. Fir begins to disappear around an elevation of 3,500 m leaving Birch as only important tree near the tree line on the southern face of Greater Himalaya. Beyond the tree line, the vegetation is found in the form of dwarfed trees, bushes, pasturelands, grasses, flowering plants etc.

3.5.4 Alpine Vegetation

The zone of alpine vegetation lies beyond the tree line and remains under snow cover during winter months. The vegetation starts blooming only after melting the snow in April. A large variety of plants including herbs, medicinal and aromatic plants, flowers, grasses, shrubs etc. are found here. It is most picturesque vegetation zone having some rare and curious varieties of plants having adoptability to high altitude harsh environment. The flowering of plants starts in April after melting the snow and continues till May. The zone has a mosaic of numerous varieties of flowering plants which are of great medicinal and aromatic values. The flowering of plants is a continuous process throughout the growing season which results into successive changes in the floristic pattern. Besides this, a major portion of this vegetation zone is covered with high altitude pastures locally called 'bugyals' which are used by Bhotias. On the basis of altitudinal variations this zone may be divided into three sub-types i.e. the lower alpine scrubland which occurs between 3,500–3,900 m with dwarf varieties of rhododendron and juniper with inter growth of bushes and scrubs, the alpine pastures lying between 3900-4400 m with alpine grasses, herbs and flowers, and the upper alpine scrubland which occurs beyond 4400 m and is the highest zone of vegetation with dry alpine scrubs.

3.5.5 Trans-Himalayan Vegetation

In this zone, there is almost absence of vegetation except some sporadic trees, bushes and pasture lands. However, due to favourable micro climate, few patches of blue pine in association with cypress, deodar, spruce, birch and high level silver Birch are found. Besides these, certain varieties of rock plants are found at high altitude dry environments.

3.6 Soils

Owing to complex physiographic conditions and high altitudinal ranges, variety of soils are found in Uttarakhand ranging from fertile alluvial soil to poor mountain soil which are scattered in small and big patches all over the state and are difficult to map to show their spatial pattern of distribution. However, these are glacial created soils, the grass land soils of high mountain, sub-mountain soil and yellow soil, foot hill soils and old alluvial soils (Raychandhuri, *et al.,* 1963).

However, the soils of Uttarakhand may be discussed under two broad heads i.e. Mountain region soils and foot hill soils.

3.6.1 Mountain Region Soils

On the basis of their process of formation these soils are grouped under two major categories i.e. Deluvial soils and alluvial soils. The deluvials are in situ soils derived from the parent materials and deposited at specific locations. These are young immature soils, generally found over the slope of the ridges and are used for agricultural activities. However, based on parent rock material, these soils are classified into Tertiary soil and older rock soil. The former generally occurs in the Siwalik ranges. These are sandy shallow porous soils which contain little humus. In limestone or dolomite zones of Siwaliks residual soils are found which are used for agriculture in the forested zone. Besides this, alluvial soils, sandy loam and clayey, are found on the southern flank of the Lesser Himalaya. The older rock soils found in the interior parts of the mountain region, have been developed from older rocks composed of slates, mica, shales and schsits and are fertile clayey with better water retention capacity. The soils derived from quartzite are light and infertile. The alluvial soils are residual soils found in low lying river valleys. These are well drained rich fertile loamy soils which are used for producing variety of agricultural crops with and without irrigation.

3.6.2 Foothill Soils Zone

On the basis of physiographic divisions, foothill soils are grouped into bhanbar soils, tarai soils and plain soils. Bhanbar soils are poor sandy soils mixed with pebbles brought down by hill streams. These are loose porous soils having very low water retention capacity. These are used for producing a variety of food and commercial crops. The tarai soils are youthful and immature with overlying layer of clay over coarse materials. The soils of low lying areas which account about 80 per cent of the total, are silty or silty loan with high calcium content. They are rich in organic and humus content and are used for cultivation of a variety of crops. On the contrary, the soils occurring on gentle slopes in the northern parts of adjoining bhanbar zone, contain high proportion of silt and fine sand but are poor in calcium content. These are well drained light grey soils with high water table and are most suitable for cultivation of rice and sugarcane. With high retention capacity and rich in humus and nitrogenous contents, these are the best fertile soils which have led to the transformation of this marshy zone to the highly mechanised scientific agriculture. The soils of the plains are, by and large, clayey with adequate organic content but are poor in calcium content. These are fertile soils used for production of variety of crops.

Soil erosion in the mountains areas and waterlogging in low lying areas specially in foot hills are the major problems of the soils in Uttarakhand.

3.7 Demographic Characteristic

According to 2011 census, the state supports a population of 10.12 million with average density of 189 person per sq. km. It has recorded 19.20 per cent decadal growth of population in 2001 over 1991 with a lot of inter and intra-district variations. In 2011 census the corresponding percentage was 19.20. The highest growth rate of nearly 33.40 per cent is recorded by US Nagar followed by Hardwar (33.16) and Dehradun (32.48), whereas lowest is found in Almora (-1.73) followed by Pauri Garhwal (-1.51) and Tehri (1.93) according to 2011 census. By and large, foot hill districts of Siwaliks have higher growth with few exceptions as compared to the rest of the state.

The spatial distribution of population in the state is influenced by a variety of landscapes together with dominance of slopes, accessibility and job opportunities. Hardwar, with only 3.6 per cent area of the state,

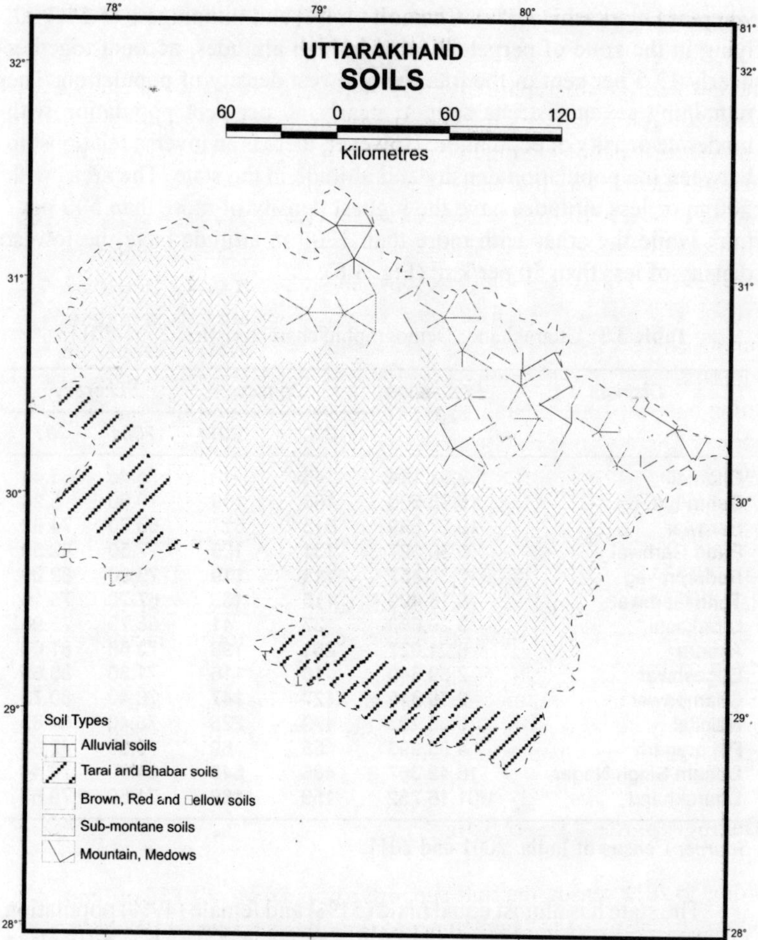

Fig. 3.6(a): Soil Map of Uttarakhand

supports 17 per cent of the total population giving highest density of 612 per/km², while Uttarkashi covering nearly 15 per cent has 3.5 per cent population with the lowest density of 37 per/km². However, it may be observed from Table 3.3 that the highest concentration of population is in foothill and plain districts and as one moves towards northeast in mountainous regions, the density of population decreases (Fig. 3.7).

The three districts of the plain viz. Hardwar (17%) US Nagar (14.6%) and Dehradun (15%) constitute nearly 47.8 per cent of the total population, whereas Uttarkashi (3.5%). Chamoli (4.4%) and Pithoragarh (5.5%) -all lying in the zone of perpetual snow in high altitudes, account together nearly 13.5 per cent of the total with lowest density of population. The remaining seven districts support nearly 60 per cent population with moderate density of population. However, there is an inverse relationship between the population density and altitude in the state. The areas with 500 m or less altitudes have the highest density of more than 500 per / km^2, while the areas with more than 1200 m attitude have the lowest density of less than 50 per/km^2 (Fig. 3.8).

Table 3.3: Uttarakhand: Demographic characteristics: 2001-2011

Districts	*Population 2011*	*Density*		*Literacy*	
		2001	*2011*	*2001*	*2011*
Chamoli	3,91,114	46	49	75.40	83.48
Dehradun	16,98,560	166	169	79.00	5.24
Hardwar	19,27,029	613	817	63.70	74.62
Pauri Garhwal	6,86,527	131	129	77.50	82.59
Rudraprayag	2,36,857	115	119	73.60	82.09
Tehri Garhwal	6,16,409	116	169	67.70	75.10
Uttarkashi	3,29,686	37	41	65.70	75.98
Almora	6,21,927	201	198	73.60	81.06
Bageshwar	2,59,840	110	116	71.30	80.69
Champawat	2,59,315	127	147	70.40	80.73
Nainital	9,55,125	179	225	78.40	84.85
Pithoragarh	4,85,993	65	69	75.90	82.93
Udham Singh Nagar	16,48,367	486	648	64.90	74.44
Uttarakhand	1,01,16,752	159	189	71.60	79.63

Source: Census of India, 2001 and 2011.

The state has almost equal male (51%) and female (49%) population but there are inter and intra-district variations in it which is difficult to generalize on regional basis. The population of the state is predominantly rural with nearly 75 per cent population living in rural areas. But there are a lot of inter and intra-district variations in it also. There are districts like Rudraprayag (98.8%) Bageshwar (96.9%), Uttarkashi (92.2%) and Tehri Garhwal (90.3%) where more than 90 per cent population lives in villages, whereas in district of Dehradun (53%) Nainital (35%) and US Nagar (32.7%) the percentage of urban population is more than the national average (2001).

The literacy level in the state is about 80 per cent which is higher than the national average (2011). The male literacy (88.3%) is higher

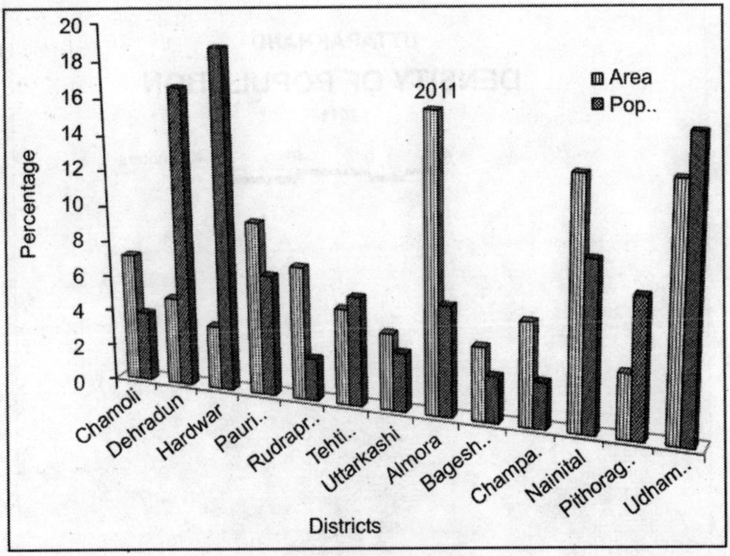

Fig. 3.7: Area and Population : 2011

than female literary (70.7%). However, there are a lot of variation in literary level too ranging from as low as 74.4 per cent in US Nagar district to an high as 85.3 per cent in Dehradun district. Leaving exceptions a part, the literacy level is higher in southeastern districts and is lower in mountain districts. The main language of the state is Hindi. Kumauni and Garhwali are the respective 'bolis' of Kumaon and Garhwal which are widely used in day to day communication by the people. However, the city culture has to a large extent swallowed up Kumauni and Garhwali and the people prefer to use Hindi instead.

The sex-ratio in the state is fairly good and in 7 out of 13 districts, female dominate the scene. In the 4 districts viz. Hardwar, Dehradun, US Nagar and Nainital sex-ratio is low which may be attributed to large influx of people migrating here in search of better living conditions by engaging in plantation or in development activities, road construction, urban expansion, infrastructure development, etc. In the rural areas the sex-ratio is much better as compared to urban centres. Similar is the case with respect to mountain districts and plain or almost plain districts.

Besides the major group of population, there are many tribal groups which are ethnically distinct but are as much part of the social environment of the region as any other cultural group. Among these, Bhotias are most

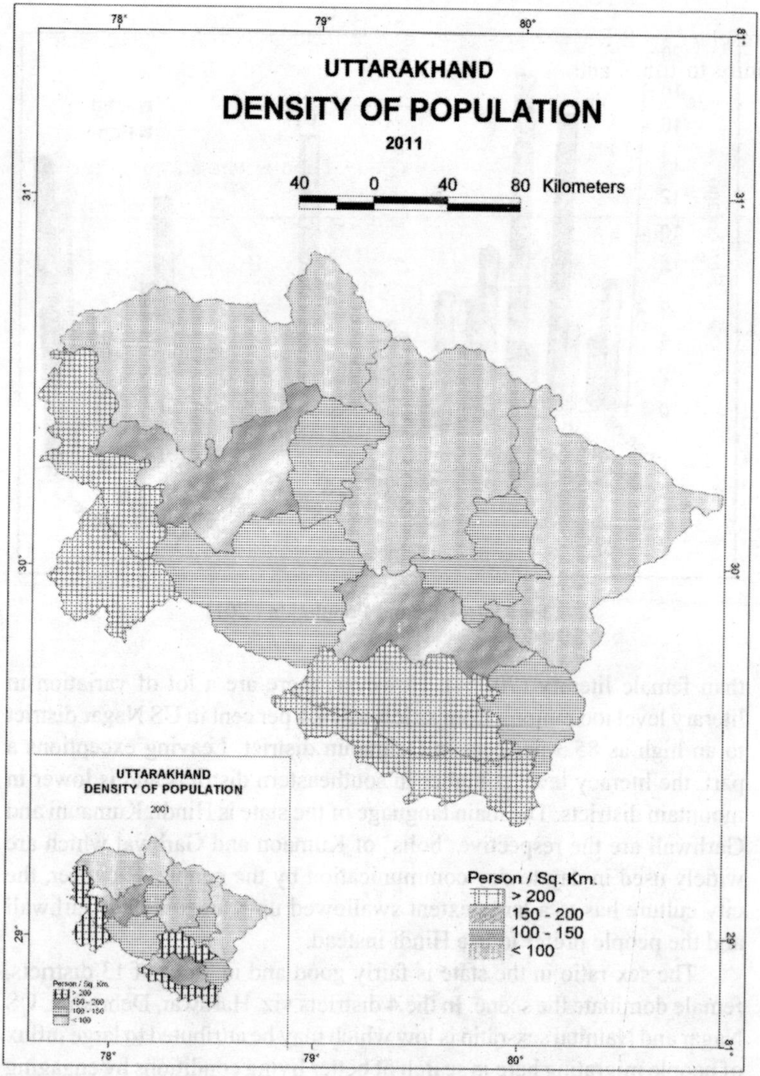

Fig. 3.8: Density of Population : 2011

prominent who inhabit a narrow belt north of the main ranges and are also known as Johari, Darnia, Byansi and Chandausi in Kumaon region and Tolchha, Marchha and Jod in Garhwal region. They are known as Jaunsari in Jaunsar Bawar region. In the tarai region. Baksha and Tharu who are often referred to as original habitants are found. Baksha associate themselves with Royal Rajput Kalan of Rajasthan but they have their affinity to tribes and belong to Mangoloid race. Like Boksha, Tharus also reveal mangoloid traits. Raji is another tribe found in Pithoragarh distinct.

The State is predominantly inhabited by Hindus who account nearly

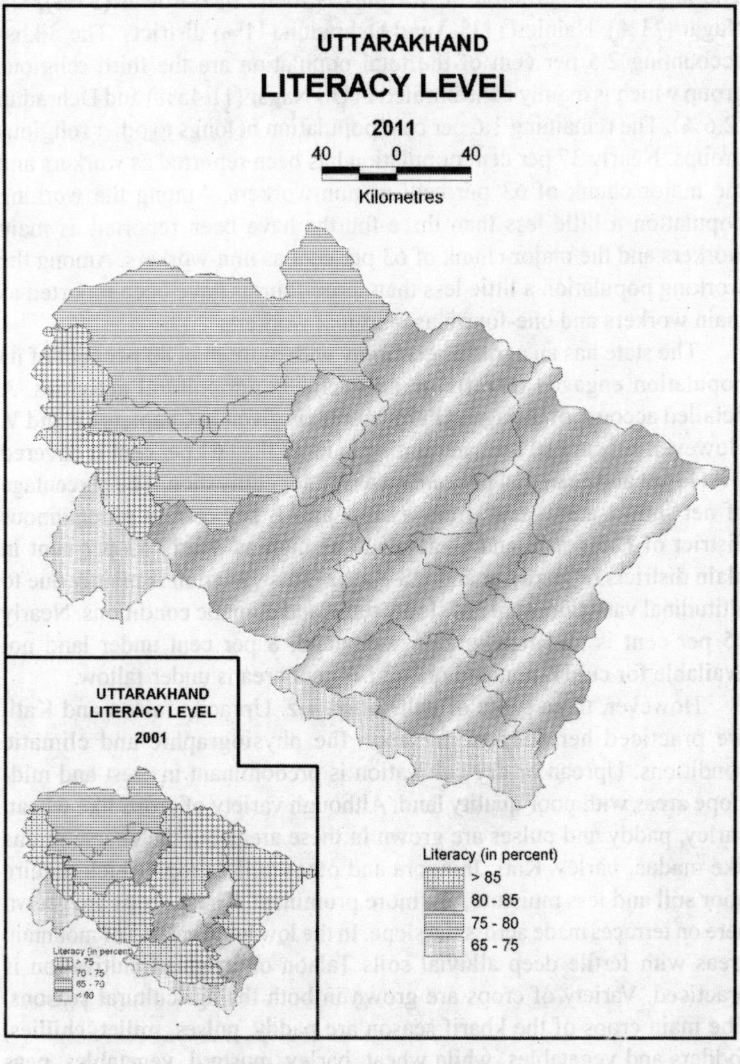

Fig. 3.9: Literacy Level-2011

85 per cent of the total population. There are 9 districts viz. Chamoli, Pauri Garhwal, Rudraprayag, Tehri Garhwal, Uttarkashi, Almora, Bageshwar, Champawat and Pithoragarh where Hindus account more than 96 per cent of the total population. Muslims–the second important religious group, constitutes nearly 12 per cent of the total population. The largest concentration of Muslims is found in Hardwar (33%), US Nagar (21%), Nainital (11%) and Dehradun (11%) districts. The Sikhs-accounting 2.5 per cent of the total population are the third religious group which is mainly concentrated in US Nagar (11.45%) and Dehradun (2.6%). The remaining 1.6 per cent population belongs to other religious groups. Nearly 37 per cent population has been reported as workers and the major chunk of 63 per cent as non-workers. Among the working population a little less than three-fourths have been reported as main workers and the major chunk of 63 per cent as non-workers. Among the working population a little less than three-fourths have been reported as main workers and one-fourth as marginal workers.

The state has an agrarian economy with more than 80 per cent of its population engaged directly or indirectly in agricultural activities. A detailed account of the agricultural system is given in Chapters IV and V. However, substantial area amounting to more than 61 per cent is covered with forest and nearly 14 per cent area is fit for cultivation. The percentage of net shown area varies from as low as 3.8 per cent in mountainous district of Uttarkashi and Chamoli to as high as nearly 53 per cent in plain districts of Hardwar and US Nagar. This variation is mainly due to altitudinal variation, angle of slope, relief and climatic conditions. Nearly 15 per cent is under culturable wasteland, 8 per cent under land not available for cultivation and only 2 per cent area is under fallow.

However, three types of cultivation viz. Upraon, Talaon and Katil are practiced here depending upon the physiographic and climatic conditions. Uproan or dry cultivation is predominant in crest and mid-slope areas with poor quality land. Although variety of crops like wheat, barley, paddy and pulses are grown in these areas but the coarse grains like madua, barley, Kati, Jhingora and other millet crops which require poor soil and less moisture, are more prominent. These crops are grown here on terraces made across the slope. In the low lying valleys of mountain areas with fertile deep alluvial soils Talaon or irrigated cultivation is practiced. Variety of crops are grown in both the agricultural seasons. The main crops of the kharif season are paddy, pulses, millet, chillies, fodders and vegetables, while wheat, barley, mustard, vegetables, peas etc. are main rabi crops. Katil or intermittent cultivation is practiced on untraced slopes which have poor soils and are unirrigated. The agriculture is slowly getting diversified. Vegetable, horticulture, seri-culture, mushroom growing, be keeping, dairy farming, pisciculture and herbal farming have better prospects of development in the state.

The state is very rich in mineral resources, which have neither yet been properly discovered nor exploited on commercial scale. However, iron, copper, marble, limestone and jewel stone are some of the important minerals found in considerable amount. Besides these, graphite, slate etc. are also found in the state. The industrial base of the state is poor except few cottage and household industries.

References

Bhakuni, K., 2000: *Identification and Development of Scenic Resources of Kumaon Himalaya,* (Unpublished Ph.D. Thesis; Department of Geography, Delhi University).

Bisht, D.S., 2001: *Guide to Garhwal and Kumaon Hills*, Trishul Pub., Dehradun.

Johnson, D.L. and Lewiss, 1995: *Land Degradation: Creation and Destruction.* Blackwell, Cambridge.

Joshi, S.C., 2004: *Uttaranchal: Environment and Development: A Geo-ecological Overview*, Gyanodaya Prakashan, Nainital.

Krishnan, M.S., 1961: *Geology of India and Burma,* Madras.

Pandey, A.C., Swagata, G. and Nathawat, M.S., 2011: *Evaluating Patterns of Temporal Glaciers in Greater Himalayan Range, J & K.*, Geo-Carto International, London.

Pant, B.R., 2000: 'Geographical Personality of Uttarakhand' in Sati, M.C. and Sati, M.C. (Eds.) *Uttaranchal State: Dimensions of Development,* Indus Pub. Co. New Delhi.

Pascoe, E.H., 1973: *A Manual of Geology of India and Burma-III,* Govt. of India Publication, Delhi.

Randhawa, H.A., 1970: *The Kumaon Himalaya*, Oxford & IBH, New Delhi.

Rawat, M.S.S., 2003: *Central Himalaya: Environment and Development* Deptt. of Geography, H.N.B. Garhwal University, Srinagar.

Rawat, A.S., 2002: *Garhwal Himalaya: A Study in Historical Perspectives,* Indus Publishing Co., N. Delhi.

Raychaudhuri, S.P., *et al.,* 1963: *Soils of India*, New Delhi.

Sati, M.C. and Sati, S.P. (Eds.), 2000: *Uttarakhand State: Dimensions of Development,* Indus Publishing Company, New Delhi.

Shah, G.L. and Rawat, A.S., 2005: A New Himalayan Entity, *Proceedings of V.C's Conference,* Nainital.

Singh, O.P., 1982: *The Himalaya- Nature, Man, and Culture,* Rajesh Publication, Delhi.

Tewari, R.C., 2006: *Geography of India*, Prayag Pustak Bhawan, Allahabad.

Valdiya, K.S., 1987: *Environmental Geology: Indian Context,* Tata McGraw Hill Pub. Co. Ltd., New Delhi.

Valdiya, K.S., 2001: Physical Divisions of Uttarakhand, *Current Science, Vol.* 81 (10).

Wadia, D.N. and Auden, J.B., 1939: Geology and Structure of Northern India, *Memoirs of Geological Survey of India*. Vol. LXXIII, Delhi.

Wadia, D.N., 1960: *Geology of India*, Mac Millan, London.

Chapter 4

Spatial Pattern of Landuse/Land Cover

4.1 Introduction

The pattern of Landuse/land cover is the function of the physio-socio-economic conditions of an area. It is more so in a mountainous state of Uttarakhand where all these conditions are highly varied and complicated. There are high mountain rang es of Greater and Trans-Himalayas with snow covered peaks, snow fields and steep slopes on the one hand, and the foothills and plains on the other hand, and in between lie the mountain ranges, ridges and river valleys with varying altitudes and degrees of slope. Likewise, it has varied and complex climate varying in almost all parametres. Similar is the case with socio-cultural and techno-economic conditions. All these varied and complicated geographical conditions have resulted into complex pattern of landuse/land cover which needs to be examined for the formulation of a rationale land use plan and policy for development to meet the growing demand of the region (Singh and Dhillon, 1984 and Shafi, 2006). It should be properly utilized, managed and conserved using modern tools and techniques (Mohammad, Singh and Dutta, 2007).

4.2 Spatial Pattern of Landuse

The general pattern of land use for the agricultural year 2005-06 is shown in Table 4.1. It would be worthwhile if a brief account of each landuse category is given here as —

4.2.1 — Forest
4.2.2 — Land not Available for Cultivation
4..2.3 — Other Uncultivable Land Excluding Fallow Land

4.2.4 — Fallow Land
4.2.5 — Net Sown Area

4.2.1 Forest

Forests are precious natural resource which not only play an important role in the regional economy but also help in pollution control and maintaining ecological balance. Forests constitute the backbone of many industries, create employment opportunities, provide fuel and fodder, protect soil erosion and regulate the flow of water through run-off and infiltration. Forest plays a crucial role in the fragile mountain ecosystem which is very sensitive to the slightest change. Therefore, they should be properly utilized, managed and conserved specially in mountain ecosystem but unfortunately it is facing a lot of problems specially deforestation, overgrazing etc. (Shafi and Raza, 1992; Singh, 1982 and Negi, 1994).

Table 4.1: Spatial Patterns of Landuse in Uttarakhand: 2005-06

Landuse categories	*Area in Hec.*	*%age of total reporting area*
1. Forest	34,65,057	61.16
2. Area not available for cultivation	4,66,674	8.24
(i) Area under non-agricultural uses	1,54,093	2.72
(ii) Barren & unculturable land	3,12,581	5.52
3. Other uncultivated land excluding fallow land	8,62,951	15.23
(i) Permanent pastures & other grazing lands	2,30,405	4.07
(ii) Land under misc. tree crops and groves not including in net area sown	2,48,934	4.39
(iii) Culturable waste land	3,83,612	6.77
4. Fallow land	1,03,514	1.82
(i) Fallow land other than current fallow	61,605	1.09
(ii) Current fallow	41,909	0.74
5. Net sown area	7,67,571	13.54
6. Total Reporting area	**56,65,767**	**100.00**

It is obvious from Table 4.1 that the state is very well endowed with very valuable natural gift of forests which occupy more than 61 per cent that was nearly 64 per cent of the total area of the state in 1992-93. The area under forest has been reported to increase to nearly 65 per cent in

2007-08 as per Forest Statistics of Uttarakhand (Forest Department, 2007-08), but the actual cover of the forest based on remote sensing and satellite imageries is reported to be only 44 per cent. This discrepancy in forest cover could be due to : (I) the forest area recorded by the Board of Revenue remains unchanged unless notified otherwise. Hence, the recorder has to record it forest even if there is not a single tree or even the land is used for cultivation of crops or other purposes; (ii) in the recent past due to insurgency in Kashmir a lot of forest area is cleared and is given to horticulture; (iii) due to increasing pressure of population a lot of forest area has been brought under farming uses; (iv) due to the government policy of afforestation specially the social forestry, a lot of area has been brought under afforestation; and (v) finally, in satellite imageries small pockets of forests are not recorded as forest.

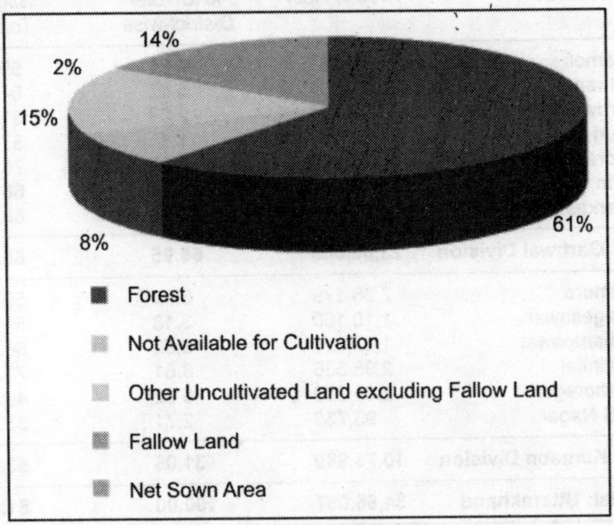

Fig. 4.1: General Landuse: 2005-06

The spatial distribution of forest is given in Table 4.2 which shows that it is not uniform across the state. The Garhwal division has more than 65 per cent of its total area under forests and it accounts about 69 per cent of total forest area, while the corresponding percentages in Kumaon division are nearly 53 and 31 per cent respectively. In Garhwal division Uttarkashi has the highest percentage (88.8%) under forest accounting nearly one-fifth of the total forest area in the state. It is followed by Rudraprayag (76.4%). These two districts together account more than one-fourth of the total forest cover in the state. These districts lie in the extreme northwestern part of the state with a vast area lying under perpetual snow and high altitude environment which is unfavourable for agriculture and human occupancy. Hence, there is lesser human intervention for deforestation. Tehri Garhwal is the third district in Garhwal division where the forest cover has exceeded the state average by more than 5 per cent. It accounts about 9.3 per cent of the forest cover of the state. Chamoli, Pauri Garhwal and Dehradun are the other districts of the division where the forest cover is less than the state average.

Table 4.2: Spatial Distribution of Forest: 2005-06

District/Division	*Area in Hec.*	*% to Total District-wise*	*%age Total*
1. Chamoli	5,06,100	14.61	59.71
2. Dehradun	2,01,831	5.82	54.71
3. Hardwar	72,431	2.09	31.35
4. Pauri Garhwal	3,85,099	11.11	57.18
5. Rudraprayag	1,79,895	5.19	76.41
6. Tehri Garhwal	3,22,051	9.29	66.30
7. Uttarkashi	7,21,661	20.83	88.83
Total: Garhwal Division	**23,89,068**	**68.95**	**65.37**
8. Almora	2,36,179	6.82	50.70
9. Bageshwar	1,10,160	3.18	51.59
10. Champawat	1,32,337	3.82	56.22
11. Nainital	2,98,336	8.61	73.40
12. Pithoragarh	2,05,239	5.92	49.58
13. US Nagar	93,738	2.71	33.57
Total: Kumaon Division	**10,75,989**	**31.06**	**53.50**
G. Total: Uttarakhand	**34,65,057**	**100.00**	**61.16**

Source: Compiled and Computed by the author.

In Kumaon division the average forest cover is about 54 per cent of its total area and it accounts 31 per cent of the total forest area of the state. In none of the districts of this division except Nainital, the forest cover is above the state average. Nainital has the highest percentage (73.4%) under forest which could be explained in terms of peoples awakening to the utility of forest cover in maintaining ecological balance and the suitability of Siwalik and Middle Himalayan ranges for afforestation. Nainital alone accounts about 9 per cent of the total forest area in the state. It is followed by Champawat where nearly 56 per cent

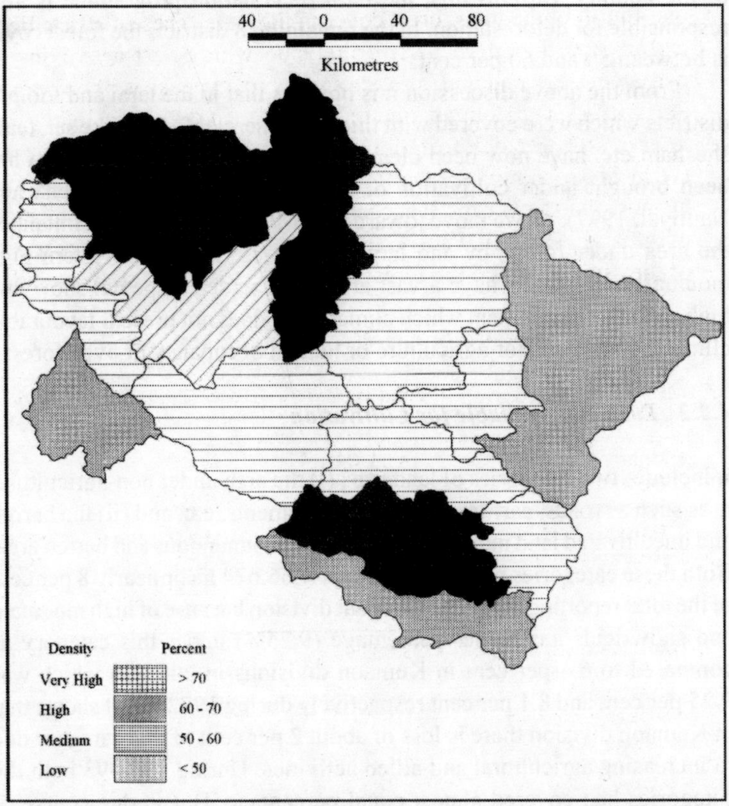

Fig. 4.2: Spatial Distribution of Forest : 2005-06

of its area is under forest. In remaining districts the forest cover is less than the divisional percentage.

The spatial distribution of forest is depicted in Fig.4.2 which shows that it is not uniform across the state. There is a contiguous patch of three districts viz. Uttarkashi, Rudraprayag and Tehri Garhwal where the forest cover is much more than the state average (61.0%). The lowest percentage is recorded in two districts viz. Hardwar and US Nagar. In Hardwar the forest cover is the lowest in the state mainly due to its location in tarai and plain areas where forests have been cleared for agricultural and allied activities. In US Nagar of Kumaon division the percentage of forest cover is second lowest in the state due to the cutting of forests for agricultural activities by Punjabi and Bengali refugees, settled here after the partition of the country. Besides this, the tarai reclamation programme is also responsible for deforestation. In the remaining 8 districts the forest cover is between 50 and 60 per cent.

From the above discussion it is obvious that in the tarai and foothill districts which were covered with thick commercial forests like sal, teak, shesham etc. have now been cleared to a large extent and the area has been brought under cultivation or other allied activities (Saxena and Nautiyal, 1997). As we move towards the Middle and Greater Himalaya, the area under forest, by and large, increases. The northwestern high mountainous districts have a vast area lying under perpetual snow and high altitude environment which climatically does not present favourable climatic conditions for agriculture or human occupancy or even forest.

4.2.2 Land not Available for Cultivation

It includes two categories of land use : (i) the area under non-agricultural uses such as roads, cart-track, canals, settlements, etc. and (ii) the barren and uncultivated land including rocky, hilly, mountainous and barren area. Both these categories put together cover 4,66,674 ha or nearly 8 per cent of the total reporting area. The Garhwal division because of high mountain and snowfields has higher percentage (9.25%) under this category as compared to 6.4 per cent in Kumaon divisions in 2005-06 which was 9.25 per cent and 8.1 per cent respectively during 1992-93. It shows that in Kumaon division there is loss of about 2 per cent of this category due to increasing agricultural and allied activities. During 1992-93 both the categories had covered almost equal percentage. But in due course of time, the areal strength of area under non-agricultural uses has declined to 2.7 per cent, while the barren and uncultivable land has recorded an increase of 5.5 per cent in 2005-06. These changing phenomena may be attributed to the abundance of settlement sites, cart tracks, loss of agricultural lands due natural hazard like landslides. However, the area under non-agricultural uses occupies nearly 12 per cent in Hardwar and 9.5 per cent in US Nagar (Udham Singh Nagar) mainly because of the expansion of settlement sites, roads, canals, cart-tracks due to increasing population. Dehradun is another district having nearly 6 per cent of its area under this category for the same reasons. In the remaining districts the area under this category is around 2 per cent or even less than this such as in Uttarkashi (0.64%) and Chamoli (0.97%). The spatial distribution of barren and uncultivable land is quite contrary to it. Chamoli has 18.75 per cent under this category, while Uttarkashi devotees nearly

5 per cent to it. The other districts having nearly 5 per cent of their land under it are Pauri Garhwal, Almora and Pithoragarh. In the remaining districts it occupies around 2 per of the total area.

It may be inferred from the above discussion that in the districts lying in Trans-Himalaya, Central Himalaya and northeastern flanks of the Siwaliks, the barren, rocky and unculturable land occupy higher percentage and the land under agricultural uses occupy less percentage mainly due to altitude, relief, slope and topography, while in the tarai foothill plain and south western flank of the Siwaliks, the condition is quite contrary. Here, the land under non-agricultural uses occupies the highest percentage, while the other category has little coverage which could be explained in terms of smooth relief, lesser slope, favourable climatic conditions for agricultural and allied activities and human occupancy. It is, therefore, suggested that more and more areas should be brought under this category without damaging the ecological balance.

4.2.3 Other Uncultivable Land Excluding Fallow Land

As is evident from Table 4.1 that this category occupies more than 15 per cent of the total area of the state, while it covers nearly 20 per cent in Kumaon division and 13 per cent in Garhwal division. During 1992-93, it had covered 14.2 per cent in the state with 11 per cent in Garhwal division and 19.3 per cent in Kumaon division. As such on the whole there are little changes in this category at state and even at divisional level over a period of 13 years. A cursory look at Appendix-V and Appendix-VI shows a lot of variations in its spatial distribution among various districts. Bageshwar district has the highest percentage (30.2%) under it followed by Pithoragarh (29.4%), Champawat (25.3%), Dehradun (21.6%) and Almora (21.6%). Hardwar (1.2%) and US Nagar (1.15%)—the plain districts, have the lowest per cent under this category with an exception of Uttarkashi (1.9%). It clearly shows that there are plenty of scope for increasing areas under agricultural and allied activities in the Shiwalik ranges and higher river valleys of central Himalaya, while in the tarai and plain areas there is little scope for it as large areas have already been brought under agricultural and allied activities.

This major category includes three sub-categories viz. (i) permanent pastures and other grazing lands; (ii) land under miscellaneous crops and groves; and (iii) culturable waste land. The areal strength of all these three categories shows a lot of variations among the districts (Appendix-V). The permanent pastures and other grazing land are popularly known as Bugyal. It had occupied about 7 per cent area of the state in 1992-93 which has declined to nearly 4 per cent in 2005-06 with 6.5 per cent in Kumaon division and 2.8 per cent in Garhwal division. In the high altitude mountain environment of northern Uttarakhand, a high proportion of pastures occur above the tree line at an elevation of more than 3,500 m, while in southern part it is included in the community pastures. The highest concentration of this category is in the districts of Champawat (11%), Bageshwar (9.5%), Pauri Garhwal (8.3%) and Pithoragarh (6.6%), while the lowest percentage is in Tehri Garhwal (0.03%), Hardwar, (6.3%), US Nagar (0.5%) and Uttarkashi (0.5) districts. The area under horticulture covers around 4.4 per cent of the total area of the state with nearly 6 per cent lying in Kumaon division and 2.5 per cent in Garhwal division. The highest percentage is recorded in Champawat (11%), Bageshwar (9.5%) and Pauri Garhwal (8.3%) where the upper river valley, spreading flanks of the mountain and suitable climatic conditions are major causes of it higher areal coverage. The districts having less than one per cent area under it are US Nagar, Uttarkashi, Hardwar and Tehri Garhwal. In Hardwar and US Nagar most of the area under it has been brought under agriculture and allied activities. In the remaining districts it covers 2 to 4 per cent of the total. However, it may be summarized that in the northern districts, where the snow covered area is the largest, the area under this category is higher due to the presence of alpine pastures, and it gradually decreases towards south. In the lesser Himalayas the occurrence of pasture is related with area under community pasture which varies from district to district. This category of landuse has the maximum potentiality to be brought under agricultural and more profitable uses.

The land under miscellaneous tree crops and groves constitutes the second category. It had occupied 4 per cent of the total area of the state in 2003-04 which has increased to 4.4 per cent in 2005-06. At divisional level nearly 6 per cent area is covered in Kumaon division because of its more suitable geographical conditions for it, while it was only 3.5 per cent in Garhwal division. A quick look at Appendix-V shows a lot of variations in its areal strength at district level. In the north, the upper reaches of the valleys of Goriganga, Ramganga, Sarju and Lodhia rivers have its higher concentration. The highest concentration in found in Bageshwar (9.5) and Champawat (11%) districts followed by Pithoragarh (6.6%) districts of Kumaon division. In Garhwal division, Pauri Garhwal accounts the highest percentage (8.3%) followed by Chamoli (4.8%) and Dehradun (4.2%). In the remaining districts its percentage is low. It may be broadly summarized that in the extreme northern mountain region and southern tarai-plain region the percentage strength of horticultural area is low, while in the central region it occupies higher percentage. However, during the recent past horticulture has emerged as a popular land use in the upland and foothill region of the state mainly due to insurgency in predominantly horticultural state of Jammu & Kashmir. Temperate fruits in the mountain areas and tropical species in narrow foothills and Dun valleys have encouraging results.

Third sub-category i.e. culturable waste land lies waste due to one or other constraints such as—altitude, topography or stony soil with low soil nutrients. Sometimes lack of irrigation facilities also gives rise to such lands. It has the maximum potentiality to be brought under cultivation and other fruitful uses. It covers only 7 per cent of total area of the state, while it occupies nearly 7 per cent in Kumaon division and 6.5 per cent in Garhwal division. Its distributional pattern is highly variable ranging from as high as 17 per cent in Dehradun to as low as 0.28 per cent in Uttarkashi. In Dehradun the higher concentration of such areas is in Vikasnagar and Rishikesh tahsils specially on the western flanks of the Siwaliks. Besides Dehradun, the other districts with higher concentration are Tehri Garhwal (16%) Pithoragarh (9.9%), Almora (9.1%) and Nainital (6.6%). In remaining districts it occupies 2 to 4 per cent. In Tehri Garhwal district it is probably associated with the fertile agricultural tract lying in the Bhagirathi valley. From the foregoing discussions it becomes obvious that in the techno-socio-culturally advanced districts its areal coverage has declined and is less because it has been brought under agricultural and other allied activities. However, it has the maximum potentiality to be brought under cultivation (Mohammad, 1992).

4.2.4 Fallow Land

This is a type of land which is left uncultivated for more than a year so that it could recoup its lost fertility due to excessive cultivation. There are three types of fallow land viz. seasonal fallow which refers to a type of land left fallow in one or the other agricultural season in the rotation system of the crops, and it is included in net sown area. The other type of fallow land is current fallow which is left unutilized in one or two successive years. The third category of fallow is old fallow which is left to recoup its fertility for 5 to 7 years; and if the agricultural land is left fallow beyond this period, it is termed as culturable wasteland. The number of years for each category is not constant and it varies from state to state.

The fallow land covers a very small percentage of the area of the state. It was only 1.36 per cent in 1992-93 which has increased to nearly 2 per cent during 2003-04 but it has again come down to 1.83 per cent in 2005-06. This minor fluctuating trend is due to either reporting error or varying doses of technological inputs.

A cursory glance at Appendix-IV shows that there is not much variation in its areal coverage at divisional level as each division has around 1.8 per cent area under it but there are considerable variations at district level. It is as high as 4.2 per cent in Champawat followed by Pauri Garhwal (3.86%) and Dehradun (3.7%) to as low as 0.27 per cent in Uttarkashi followed by Chamoli (0.3%) and Bageshwar (0.4%). In the remaining districts it varies from one to two per cent.

This is a type of land which has the maximum potentiality to be brought under cultivation with minimum efforts of putting more compost manure or chemical fertilizer and proper management of the land resources through contour farming, bunding and maintaining the terraces properly. It would ultimately increase the areal strength of agricultural land which will in turn increase agricultural production.

4.2.5 Net Sown Area

This is the real agricultural land on which the farmers perform all their agricultural activities to get their livelihood. The farmers always try their level best to bring more and more land under cultivation but in Uttarakhand it has shown a very slow and steady progress in it because the level land is scarce and is available in wide river valleys which are very few and far between. It is also found in certain ridges and mid-crest zones where the nature of slope, structure of soil, water resources and sunshine are favourable. Altitude is also a determining factor in the mountainous area, and on an average the largest portion of cultivated land falls between the elevation range of 750 to 1500 m. The highest percentage of net sown area is found in foothills, plains and Dun valleys. It was nearly 12.4 per cent during 1992-93 which has increase to about 13.4 per cent in 2004-05. In 2005-06 it has recorded a minor growth of 0.13 per cent in its area over 2003-04. In the light of growing demand of agricultural land, this slow and steady growth in it may be explained in two ways : Firstly, although more and more land is brought under agriculture on the one hand, but on the other hand, the agricultural land is given to non-agricultural uses such as settlements, canals, roads etc., and secondly, due to movement of working population specially the educated persons to the plain in search of better job opportunities, the land is, in most cases, abandoned from agricultural uses. However, the district-wise areal strength of the net sown area is given in Table 4.3. Moreover, the net sown area is also shrinking due to its unscientific use, over utilization and mismanagement of land resources (Patra, 2006 and Thakur, 2011).

Table 4.3: Spatial Patterns of Distribution of Net Sown Area: 2005-06

	District/Division	*Area in Hectare*	*% to total area of the district*	*% to N.S.A. of State*
1.	Chamoli	32,560	3.84	4.24
2.	Dehradun	48,508	13.15	6.32
3.	Hardwar	1,20,581	52.18	15.71
4.	Pauri Garhwal	81,491	12.10	10.62
5.	Rudraprayag	19,516	8.29	2.54
6.	Tehri Garhwal	59,120	12.17	7.70
7.	Uttarkashi	30,819	3.79	4.02
Total: Garhwal Division		**3,92,595**	**10.74**	**51.15**
8.	Almora	80,951	17.38	10.55
9.	Bageshwar	26,567	12.44	3.46
10.	Champawat	23,965	10.18	3.12
11.	Nainital	47,730	11.74	6.22
12.	Pithoragarh	45,902	11.18	5.98
13.	US Nagar	1,49,861	53.67	19.52
Total: Kumaon Division		**3,74,976**	**18.64**	**48.85**
G. Total: Uttarakhand		**7,67,571**	**13.55**	**100.00**

Source: Compiled and Computed by the author.

It is obvious from Table 4.3 that there are a lot of variations in the areal strength of net sown area. The Kumaon division has nearly 19 per cent of its area under this category, while it is only 11 per cent in Garhwal division. The higher percentage in Kumaon division is mainly due to highest percentage in US Nagar (53.7%) which is the most advanced districts in agricultural technology. Consequently, more areas have been brought under cultivation. The other districts of this division have also comparatively higher percentage under this category such as Almora (17.4%), Nainital (11.7%) and Bageshwar (12.4%). In all these districts lower broad river valleys and favourable climatic conditions are the main contributing factors in this respect. In Almora district the valleys of Kosi and Ramganga, have larger area under this category, while Bageshwar district is gifted with extensive tracts of fertile land in Gomati and Sarju river basins. In Nainital district the foothill bhanbar tract of Bhowali-Bhimtal valley and Kalsa Valley contribute more to net sown area, while

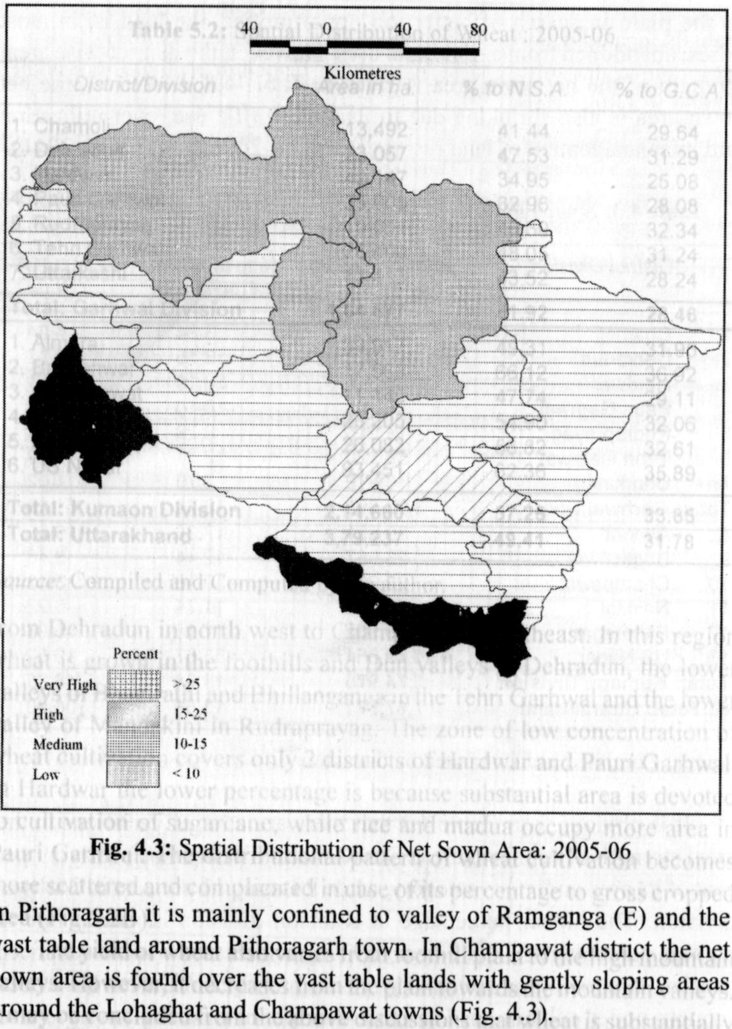

Fig. 4.3: Spatial Distribution of Net Sown Area: 2005-06

in Pithoragarh it is mainly confined to valley of Ramganga (E) and the vast table land around Pithoragarh town. In Champawat district the net sown area is found over the vast table lands with gently sloping areas around the Lohaghat and Champawat towns (Fig. 4.3).

In Garhwal division the highest percentage (52.2%) of net sown area is found in Hardwar district which is located in the Upper Ganga Plain. It is smooth fertile level plain having favourable climatic and socio-economic conditions for agricultural activities. The other districts having reasonably higher percentage under this category are Dehradun (13.2%), Tehri Garhwal (12.2%), and Pauri Garhwal (12.1%). The higher areal coverage in Dehradun is due to fertile plain area of Vikasnagar and Rishikesh tahsils. In Pauri Garhwal district Kotdwar, and Landsdown tahsils located in fertile foothill plains, have contributed to higher areal strength of net sown area. The lowest percentage of net sown area is found in Uttarkashi (3.8%) and Chamoli (3.84%) which are located in high mountainous area having unsuitable relief, altitude, and climatic conditions for agricultural activities. Rudraprayag has also high percentage (8.3%) because of broader river valleys and outwash plains (Singh, 1965).

One the whole Uttarakhand, being a mountainous state has only 13.6 per cent of its total area under cultivation which is far below the national average (46%) but it has a huge potentiality of expanding agricultural land on the less sloping sun facing slopes having suitable soil and climatic conditions, and the lower fertile valleys of the innumerable rivers of the states. But unfortunately the younger generation of the state does not want to venture in agricultural pursuits and are migrating towards the urban centres in the plain in search of better job option in order to lead a more healthy and comfortable life. The government should, therefore, formulate agricultural programmes and policies for bringing more land under agriculture and also provide various kinds of incentives to the younger generation of the state who does not want to venture in agricultural pursuits and are migrating towards the urban centres in the plain in search of better job opportunities in order to lead a more healthy and prosperous life. The government should, therefore, formulate agricultural programmes and policies for bringing more land under agriculture and also provide various kinds of incentives to the younger generation in the form of allied activities to control their outward migration.

The Spatial distribution of net sown area is represented in Fig. 4.3 which shows that there are altogether 4 districts viz. Hardwar, Dehradun, Almora and US Nagar where the percentage of net sown area is more than the state average. Dehradun and Hardwar form a contiguous belt, while Almora and US Nagar are separated by Nainital. The 3 districts viz. Uttarkashi, Rudraprayag and Chamoli forming a contiguous belt in northwest mountainous region of the state have low percentage of net sown area. In remaining 6 districts, it ranges between 10 and 13 per cent. On the whole, it may be concluded that the spatial distribution of agricultural land is largely an outcome of the complex interplay of various physical factors together with certain socio-cultural factors right from the high altitude environment close to snow ranges to the subtropical zone of plain and tarai in the extreme south of the state (Fig. 4.3).

4.3 Spatial Pattern of Irrigation

Water is prime input to agriculture whether it is available through precipitation or irrigation because even the metabolic process will not begin without it what to say of vigorous vegetative growth and higher yield of crops. Since, like our country, in Uttarakhand also the precipitation is periodic, seasonal and highly variable in its time of incidence, retreat, amount of rainfall and its intensity. All these uncertainties in the occurrence of precipitation necessitate the use of irrigation water. Scientific and proper utilization of water resources for irrigation will enhance agricultural productivity which will improve the status of food security (Narasaiah, 2003).

Table 4.4: Sources of Irrigation: 2005-06

Sources	Garhwal Division		Kumaon Division		Uttarakhand	
	Area	%age	Area	%age	Area	%age
Wells	93,994	60.82	1,19,586	63.32	2,13,580	62.19
Canals	38,604	24.98	57,004	30.18	95,608	27.84
Others	21,952	14.20	12,272	6.50	34,224	9.97
Total	**1,54,550**	**100.00**	**1,88,862**	**100.00**	**3,43,412**	**100.00**

Source: Compiled and Computed by the author.

Owing to a large number of seasonal and perennial rivers and adequate amount of precipitation, Uttarakhand has huge potential of irrigation but it has not been rationally utilized. Nearly 44.8 per cent of the net sown area is irrigated in the state. In Garhwal division because of colder climate and seepage of water, only 36.8 per cent of the net sown area is irrigated, while in Kumaon division where climate is warm, water table is comparatively high, the percentage of irrigated land is nearly half of the net sown area. There are two main sources of irrigation i.e. canals and wells. Besides these, there are other sources like fountain, chu, small rivulets, kuls, etc. which have been put under other sources. The areal strength together its percentage to total is given in Table 4.3 which shows that wells including tube wells and other well are the predominant source of irrigation and account more than 62 per cent of the irrigated area in the state. It is followed by canals which accounts nearly 28 per cent of irrigated area. About one-tenth of net sown area is irrigated through several other sources.

There are division-wise differences in these sources of irrigation. In Kumaon division where the boring of well is easy and less costly due to higher level of water table and thick soft geological rock structure, tube-wells account nearly 79 per cent of the irrigated area, while in Garhwal

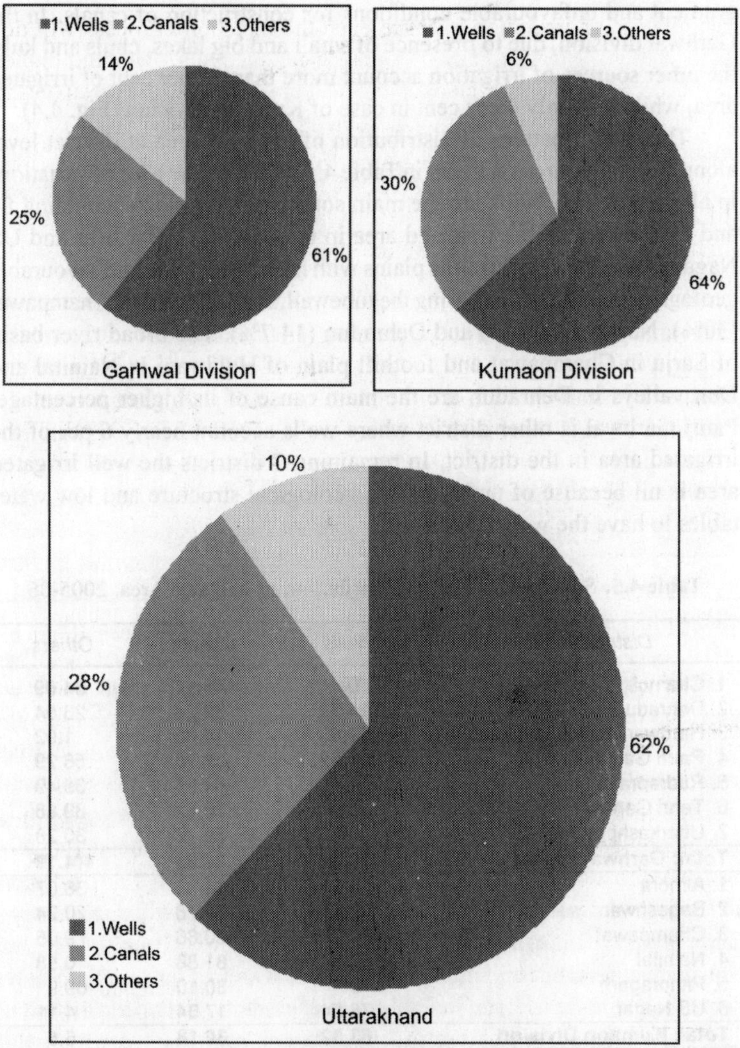

Fig. 4.4: Sources of Irrigation : 2005-06

division because of lesser favourable conditions it accounts nearly 61 per cent. But, due to higher gradient facilitating gravity flow of water and presence of large number of perennial and non-perennial streams, canals account nearly one-fourth of the irrigated area in Garhwal division, while its contribution is only 17 per cent in Kumaon division because of lower gradient and unfavourable conditions for construction of canals. In the Garhwal division, due to presence of small and big lakes, chuls and kuls, the other sources of irrigation account more than 14 per cent of irrigated area, while it is only 4 per cent in case of Kumaon division (Fig. 4.4).

The spatial pattern of distribution of irrigated area at district level along with the sources is given in Table 4.5 which shows a lot of variations in all parameters. Wells are the main source of irrigation accounting 84 and 79 per cent of the irrigated area in the district of Hardwar and US Nagar—both lying in foothills plains with high water table and favourable geological conditions for boring the tubewell. It is followed by Champawat (30%), Nainital (17.5%) and Dehradun (14.7%). The broad river basin of Sarju in Champawat and foothill plain of Haldwani in Nainital and Dun valleys in Dehradun are the main cause of its higher percentage. Pauri Garhwal is other district where wells account nearly 6 per of the irrigated area in the district. In remaining 7 districts the well irrigated area is nil because of unfavourable geological structure and low water tables to have the wells (Table 4.5).

Table 4.5: Source-wise Spatial Distribution of Irrigated Area: 2005-06

District/Division	*Wells*	*Canals*	*Others*
1. Chamoli	0	45.91	54.09
2. Dehradun	14.72	62.24	23.04
3. Hardwar	84.03	14.05	1.92
4. Pauri Garhwal	5.52	37.78	56.29
5. Rudraprayag	0	64.51	35.49
6. Tehri Garhwal	0	10.12	89.88
7. Uttarkashi	0	69.72	30.28
Total: Garhwal Division	**60.82**	**24.98**	**14.19**
1. Almora	0	63.33	36.67
2. Bageshwar	0	79.76	20.24
3. Champawat	30.09	60.86	9.05
4. Nainital	17.54	81.88	0.58
5. Pithoragarh	0	30.10	69.9
6. US Nagar	78.54	17.34	4.11
Total: Kumaon Division	**63.32**	**30.18**	**6.5**
Total: Uttarakhand	**62.19**	**27.84**	**9.97**

Source: Compiled and Computed by the author.

Canal is a popular source of irrigation, although it irrigates only 27.8 per cent of net irrigated area. It is popular source of irrigation practiced in all the districts because of the plenty of surface water and favourable gradient for downward flow of water. It ranges from 10 per cent in Tehri

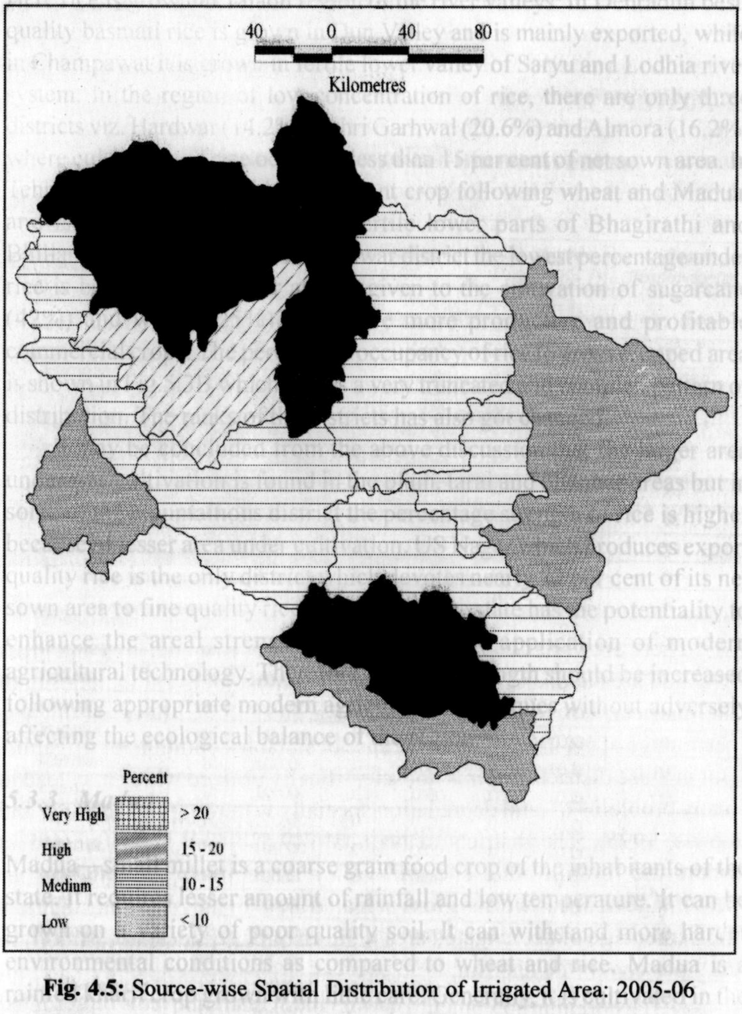

Fig. 4.5: Source-wise Spatial Distribution of Irrigated Area: 2005-06

Garhwal to as high as 82 per cent in Nainital. There are six districts viz. Bageshwar (79.8%), Uttarkashi (69.7%), Rudraprayag (64.5%), Dehradun (62.2%) and Champawat where canal irrigated area accounts more than 60 per cent of the total irrigated area of the respective districts. In remaining 6 districts the canal irrigated area is high. Other sources of irrigation have accounted higher percentages of irrigated area, whereascanals and wells have contributed lesser percentage. These are mostly high mountainous districts.

Photo 4.1: A view of Tehri Dam

There are variation in the percentage of net irrigated area and gross irrigated area at district level but they are almost identical in majority of the districts. US Nagar (95.7%) and Hardwar (87%) have the highest percentage of net irrigated area because of suitable condition for irrigation. But in these districts water is not scientifically utilized which may lead to many problems of land degradation specially waterlogging (Johnson and Lewis, 1995). The districts of Nainital (49.5%) and Dehradun (45%) where the foothill plain and dun valley respectively have the higher percentage of net as well as gross irrigated area. In the remaining districts the percentage of net irrigated area varies from 7.8 per cent (Chamoli) to 23.9 per cent in Bageshwar.

The spatial distribution of irrigated area at district level is represented in Fig. 4.5 which shows that the percentage of net irrigated area to net sown area is more than 45 per cent in 4 districts forming two clusters of two districts each viz. Dehradun and Hardwar, and Nainital and US Nagar. The districts of low intensity of irrigation with less than 15 per cent irrigated area are also found in two pockets including 4 districts. The largest pocket including the districts of Uttarkashi, Tehri Garhwal and Rudraprayag is found in northwestern high mountain region of the state, while the other pocket includes only Bageshwar district. In the remaining

5 districts the percentage ranges from 15 to 45. They form a continuous region with the only interruption made by Bageshwar district (Fig. 4.5).

Since the state has huge potential of water resources in large number of small and big seasonal and perennial stream, it should be properly and rationally utilized so that not only agricultural development takes place but agricultural practices, cropping pattern and agricultural activities may be scientifically diversified.

References

Agricultural Statistics at a Glance : 2005 to 2010, Directorate of Economics and Statistics, Ministry of Agriculture, Government of India, New Delhi.

Indian Agricultural Statistics, Vol. I & II, Directorate of Economics and Statistics, Ministry of Agriculture, Government of India, New Delhi.

Johnson, D.L. and Lewis, L.A., 1995 : *Land Degradation : Creation and Destruction*, Blackwell, Cambridge.

Mohammad, N., 1988 : 'Wasteland in India—Some Issues and Problems', in Sharma, S.C. (Ed.), *Utilization of Wasteland for Sustainable Development*, Concept Publication Co., New Delhi.

Mohammad, N., Singh, R.B. and Dutta, A. (Eds.), 2007 : *Spatial Information Technology for Natural Resource Management*, Concept Pub. Co., New Delhi.

Narasaiah, M.L., 2003 : *Food Security and Irrigation*, Discovery Pub. House, New Delhi.

Negi, G.C.S., 1994 : Agro- forestry System in Central Himalayan Mountains of India, *Agro forestry Today*.

Patra, P., 2006 : *Land Degradation in Tropical Highlands*, Rajesh Pub., New Delhi.

Saxena, A.K. and Nautiyal, J.C., 1997. Analysing Deforestation : A Dynamic System Approach, *Journal of Sustainable Forestry*, Vol.V, Dehradun.

Shafi, M. and Raza, M. (Eds.), 1992 : *Forest Ecosystem of the World*, Rawat Pub., Jaipur.

Shafi, M., 2006 : *Agricultural Geography*, Pearson Education in South Asia, New Delhi.

Singh, J. and Dhillon, S.S., 1984 : *Agricultural Geography*, Tata McGraw Hill Pub., New Delhi.

Singh, J.S. (Ed.), 1985 : *Environmental Regeneration in Himalaya : Concept and Strategies*, Gyanendra Prakashan, Nainital.

Singh, L.R., 1965 : *The Tarai Region of U.P.*, Allahabad.

Singh, O.P. 1982 : *The Himalaya : Nature, Man, and Culture*, Rajesh Pub., New Delhi.

Thakur, B. 2011 : *Depletion and Degradation of Land Resources of India*, R.N. Dubey Foundation, Allahabad.

Uttarakhand: Forest Statistics (2007-08), Forest Department, Government of Uttarakhand, Dehradun.

Chapter 5

Dynamics of Agricultural Landuse and its Diversification

5.1 Introduction

In spite of the limited land available for cultivation, agriculture remains the basic activity in Uttarakhand engaging nearly 80 per cent of its total population directly or otherwise. Like any other region, here also agriculture is the manifestation of the interplay of combined set of physical, techno-economic and socio-cultural factors with their varying degree of influence. In the mountainous state physical factors surface configuration, altitude, angle of slope and its aspect, climatic parameters specially the temperature and precipitation, water resources and soil play more dominant role as compared to other set of factors. Here topography highly restricts the agricultural activity as the level land is very scarce and is available only in river valleys which are very few and far between. The agricultural area also extends to the ridges and mid-crest zones wherever the nature of slope, soil structure and texture, irrigation water and bright sunshine are favourable. However, in most of the mountain regions of Uttarakhand the largest proportion of agricultural land lies between 750-1500 m. In the tarai and foothills region with an average elevation of 500 m. a large tract of land is devoted to agriculture because of favourable environmental conditions (Singh, 1985; Masood, 2000, and Nag, 2010).

Thus it is clear that according to the variations in surface configuration and climatic conditions there are two broad environmental types; viz. the mountainous zone and foothill plains. The mountainous zone is the largest zone covering the entire state north of the foothills. It has a wide range of topographical and climatic conditions leading to variations in farming system and crops which is also influenced by local factors of soil, availability of water, slope, aspects etc. According to the altitudinal range this broad zone is divided into 4 subzones viz. the low river valleys, crest and mid-slopes, high barren slopes and the high altitude crop zone (Fig. 5.1).

The low river valleys, lying between the altitudinal ranges of 1000 to 1500 m mainly occur in Lesser Himalayan zone. They are characterized by level plain, deep soil, plenty of irrigation water, and favourable climatic parameters. The agricultural type practiced here is known as Talaon agriculture or low land agriculture. Variety of crops ranging from rice and wheat in the lower area with deep soil and plenty of water to maize, madua and other millet in the higher areas, are produced here with high yield per unit of land. Thus these valleys are the granaries of the mountainous Uttarakhand. The crest and mid-slopes are intermediate elevations lying between 1500 to 2000 m with dominance of sloping areas. It has a thin soil cover of varying fertility, and temperature with increasing heights which reduces the growing period. Here terrace farming is practiced which provides level plain and prevents soil erosion. Terraces of various sizes are found over the vast areas at various altitudes in the middle Uttarakhand. Several food crops of both kharif and rabi seasons are grown here. Leaving limited area where irrigation farming is practiced, dry farming dependent on the vagaries of weather is practiced in these areas. The yield of crops is obviously low. The type of agriculture practiced here is known as Uproar—the cultivation on upper elevations.

The high barren slopes lying between 2000 to 2500 m are characterized by thin cover of infertile soil, barren land, no irrigation facilities and low temperature for most part of the year. Constrained with increasing demand of foodgrains, the farmers resort to a primitive type of farming called Katil wherein agriculture is practiced without irrigation depending upon the mercy of rain god and even without terracing the slopes. Certain varieties of coarse grains and poor quality wheat are produced here with very low yield due to vagaries of environmental conditions (Joshi, 2004 and Maikhuri *et al.*, 2004).

The high attitude crop zone lying in the northern borders roughly above 3000 m altitude, are mostly snow covered areas with thin layer of soil on slopes and harsh climatic condition providing limited period of crop cultivation. The agricultural operation begins in the month of March when snow melts with increase in temperature. Here, agriculture is not the basic activity and is carried out depending upon the availability of facilities. The area is inhabited by pastoral tribal population resorting to seasonal migration according to climatic conditions. They do not rely much on agriculture and supplement their economy with cottage industries and trade.

The foothill plains constitute the bhanbar foothills, tarai and plains. The bhanbar foothills is a narrow belt lying between tarai in the south and the mountain areas in the north mainly in the districts of Nainital and Dehradun. In Nainital it has undulating relief with light gradient which requires the creation of lèvel terraced fields for cultivation and also to prevent soil erosion. The soil is deep but porous and water table is low. Hence, irrigation farming is practiced and 3 crops are grown in a year during kharif, rabi and zaid seasons. In Dehradun, the Dun Valley has similar conditions of relief and soil depth but there is no scarcity of surface water, and the rabi crops and now even the kharif crops are produced with the help of irrigation. The crops are produced all the year round during kharif, rabi and zaid seasons. In the recent past the greater emphasis is given to the growing of cash crops like sugarcane, vegetables, basmati rice, soyabean and vegetables.

The Tarai-Plain area consists of eastern tarai zone and western plain area. The former covers the whole of US Nagar and is characterized by almost level plain, deep fertile soil, high water table and favourable climatic conditions which provide suitable base for mechanization and commercialization of agriculture. Rice, sugarcane and wheat are the principal crops grown here. The western plain covers whole of Hardwar district and Kashipur development block of US Nagar. It is similar to tarai zone but does not have abundance of surface water. A large variety of food crops, fibres, oil seeds are grown with the help of irrigation in this tract (Jodha, 1992).

5.2 Agricultural Practices

Owing to varied ecological, techno-economic and socio-cultural factors different types of agricultural systems varying from intensive to extension, subsistence to commercial, primitive to modernized are practical here like many other parts of the country. But here a special type of cultivation is practiced which are locally known as Uproar, Talaon and Katil (Bandooni, 2004).

The Uproar also known as upland or dry cultivation is mostly practiced without irrigation in the crest and mid-slope areas where thin layer of infertile soil is found. Here, narrow terraces scattered in pockets are made across the slope which helps in absorption of rain water in the soil increasing groundwater level and also protecting the soil from erosion. These terraces are to be constantly maintained year after year. Variety of foodgrains like wheat, rice, barley, pulses, madua, Jhingora etc. are grown

Photo 5.1: GBP Agricultural University, Pant Nagar

but coarse grains are more prominent. The land is to be left follow in one season within a period of two years owing to the infertility of soil. The standard practice in Uttarakhand is that the land is put to cultivation in two kharif seasons and one rabi season and in the next rabi season land is left fallow. Within this system also there are variations at micro-level.

The Talaon also called as low land or irrigated cultivation is generally practiced in the low lying areas of the important mountain river valleys specially in Bhagirathi, Bhillanganga, Alaknanda, Ramganga, Kosi, Sarju, Yamini, Kali, etc. These areas are characterized by deep fertile alluvial soils with high moisture absorption capacity, high humus content and high moisture holding capacity, high temperature condition and abundance of irrigation water. Since, these valleys do not have large level plains, terracing is practiced with greater width of terraces. Canal irrigation is common practice in these areas for successful cultivation. These are the areas of intensive cultivation and variety of crops is grown during both kharif and rabi seasons. The major crops of kharif season are paddy, maize, madua, jhingora, pulses, chillies, fodder, etc., while the principal crops of rabi season are wheat, barley, potato, peas, oil seeds, vegetables etc. Large amount of foodgrains specially rice is grown in these low lying mountainous valleys. Katil also known as intermittent cultivation is usually practiced on unirrigated upland area with such a poor soil that the land had to be left-follow for one or two years to recoup its fertility. The crops are grown on unterraced slope which cause excessive run off and soil erosion. The crop yield is poor and the farmers are abandoning this practice as it is uneconomical (Bandooni, 2004 and Alam, 2003).

5.3 Spatial Pattern of Agricultural Landuse

Since all the geographical factors- physio-ecological, techno-economic and socio-cultural are highly varied and complex at all levels from macro to micro both in space and time on the one hand, and the requirement of these geographical factors varies for all the species of agricultural plants right from the process of metabolism to the ripening of crops, there are a lot of variations in the cultivation of crops across the state (Mohammad, 1972 and Ahmad, 2004). Variety of crops is grown in the state but the areal strength together with the percentage of some of the crops is given in Table 5.1.

Table 5.1: Area under Principal Crops in Uttarakhand

Principal Crops	*2003-04*		*2005-06*	
	Area in ha	*%age to total*	*Area in ha*	*%age to total*
1. Wheat	3,83,385	31.39	3,79,237	31.28
2. Rice	2,77,326	22.70	2,84,430	23.46
3. Madua	1,26,423	10.35	1,24,369	10.26
4. Vegetables	31,147	2.71	34,802	2.87
5. Maize	31,346	2.57	31,461	2.60
6. Barley	24,940	2.04	23,420	1.93
7. Pulses	50,568	4.14	51,560	4.25
8. Oilseeds	30,474	2.49	34,043	2.81
9. Sugarcane	1,16,877	9.57	1,07,748	8.89
10. Condiments	9,140	0.75	9,015	0.74
11. Fruits	15,131	1.24	15,886	1.31
Drugs & Narcotics	1,199	0.10	923	0.08
Others	1,12,769	9.95	1,15,473	9.52
Total	**12,21,510**	**100.00**	**12,12,367**	**100.00**

Source : Compiled and Computed by the author.

A cursory glance at Table 5.1 shows that almost all the crops have recorded increase in their percentage strength to gross cropped area in 2005-06 over 2003-04 but the quantum of growth is very less. Hence, the analysis is confined to the agricultural year 2005-06. Table 5.1 further shows that cereals and millets account nearly 70 per cent of gross cropped area of the state. They cover 77 per cent in Kumaon division and 61 per cent in Garhwal division. Rice and wheat accounting together nearly 55 per cent are the predominant cereals in the state followed by madua which occupies about one-tenth of the gross cropped area. But these two crops occupy nearly 64 per cent in Kumaon division and 45 per cent in Garhwal division. Sugarcane is the other important crop which occupies nearly 9 per cent at state level and 13 and 5 per cent in Garhwal and Kumaon

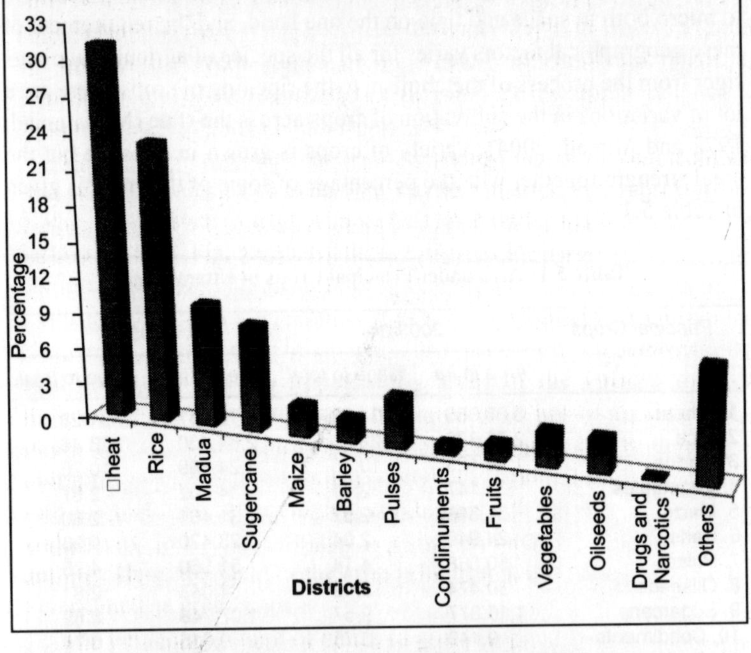

Fig. 5.1: Area under Principal Crops : 2005-06

divisions respectively. The other crops covering significant percentage are pulses, vegetables and oilseeds which together occupy nearly one-tenth of the gross cropped area in the state. On the whole food crops put together occupy nearly 85 per cent of gross cropped area in Uttarakhand, while in Kumaon and Garhwal divisions they cover nearly 91 and 84 per cent of gross cropped area respectively.

The above discussion highlights the aggregate pattern of spatial distribution of crops in the state but the data given in Appendix-VIII shows that there are a lot of inter and intra-districts variations in the percentage strength of the different crops at district level. Therefore, it would be worthwhile to discuss the spatial distribution of major crops in little detail.

5.3.1 Wheat

Wheat, occupying more than 41 per cent of the gross cropped area is a predominant crop of the state. It is a crop of rabi season with varying season of growth according to altitudinal variations. In the foothill and plains it is sown in November-December and is harvested in March-April allowing a third crops in summer season. But ascending towards higher mountain valleys, the growing season starts increasing and around 3000 m elevation its cultivation begins in mid-March after the melting of snow, and it takes nearly 4 months to ripen resulting into the loss of summer crops.

The areal strength of wheat has continuously declined over the last 15 years because of agricultural diversification. In 1992-93, 4,09,037 ha amounting to 52.2 per cent of net sown area and 32 per cent of gross cropped area were devoted to wheat cultivation, while in 2003-04 it has declined to 3,83,386 ha amounting to 50.4 per cent of net sown area and 31.3 per cent of gross cropped area. During 2005-06 it has further declined to 3,79,237 ha contributing 49.4 per cent to net sown area and 31.3 per cent to gross cropped area (Appendix-IX). However, there are inter and intra-district variations in the percentage strength of wheat.

The areal strength of wheat in different districts in given in Table 5.2 and its distribution is shown in Fig. 5.2 for the year 2005-06. Fig.5.2 A shows that there are 4 districts viz. US Nagar (64.2%), Nainital (54.9%), Bageshwar (66.1%) and Pithoragarh (56.8%) where wheat occupies more than 50 per cent of net sown area. In US Nagar the fertile alluvial soil, favourable climatic condition and modern agricultural technology together with enterprising Punjabi farmers having adequate knowledge of technique of wheat cultivation, are the main causes of high area under wheat. In Nainital district it is grown in foothills and plain areas of Haldwani tahsil and low land of Bhowali area. The other sub-region of high concentration of wheat cultivation consists of Bageshwar and Pithoragarh districts. In this zone its cultivation is confined to its eastern part in the river valleys of Dhauliganga, Goriganga, Ramganga and Saryu where the fertile valley soil, warm climatic condition and better irrigation facilities provide conducive conditions for its cultivation.

The region of moderate concentration with 40-50 per cent of net sown area spreads over 7 districts. It forms a contiguous region extending

Table 5.2: Spatial Distribution of Wheat : 2005-06

District/Division	Area in ha.	% to N.S.A.	% to G.C.A.
1. Chamoli	13,492	41.44	29.64
2. Dehradun	23,057	47.53	31.29
3. Hardwar	42,147	34.95	25.08
4. Pauri Garhwal	35,005	32.96	28.08
5. Rudraprayag	9,054	46.39	32.34
6. Tehri Garhwal	28,406	48.05	31.24
7. Uttarkashi	13,413	43.52	28.24
Total: Garhwal Division	**1,64,577**	**41.92**	**28.46**
1. Almora	39,917	49.31	31.95
2. Bageshwar	17,565	66.12	36.92
3. Champawat	11,440	47.74	29.11
4. Nainital	26,205	54.90	32.06
5. Pithoragarh	26,082	56.82	32.61
6. US Nagar	93,451	62.36	35.89
Total: Kumaon Division	**2,14,660**	**57.25**	**33.85**
Total: Uttarakhand	**3,79,237**	**49.41**	**31.78**

Source: Compiled and Computed by the author.

from Dehradun in north west to Champawat in southeast. In this region wheat is grown in the foothills and Dun valleys in Dehradun, the lower valleys of Bhagirathi and Bhillanganga in the Tehri Garhwal and the lower valley of Mandakini in Rudraprayag. The zone of low concentration of wheat cultivation covers only 2 districts of Hardwar and Pauri Garhwal. In Hardwar the lower percentage is because substantial area is devoted to cultivation of sugarcane, while rice and madua occupy more area in Pauri Garhwal. The distributional pattern of wheat cultivation becomes more scattered and complicated in case of its percentage to gross cropped area (Fig.5.2B).

The yield of wheat also varies from foothill plain to the high mountain valleys. However, it decreases from the plain towards the mountain valleys. It may be concluded from the above discussions that wheat is substantially grown in talaon besides the foothills and plain areas, and a small area is found in Upraon and almost mid-Katil areas. There are plenty of scopes for increasing the areal strength and the production of wheat in Uttarakhand by adopting new techniques of farming.

5.3.2 Rice

Rice is the next important crop of the state in terms of its areal strength and is the main staple crop of its inhabitants. It requires plenty of water,

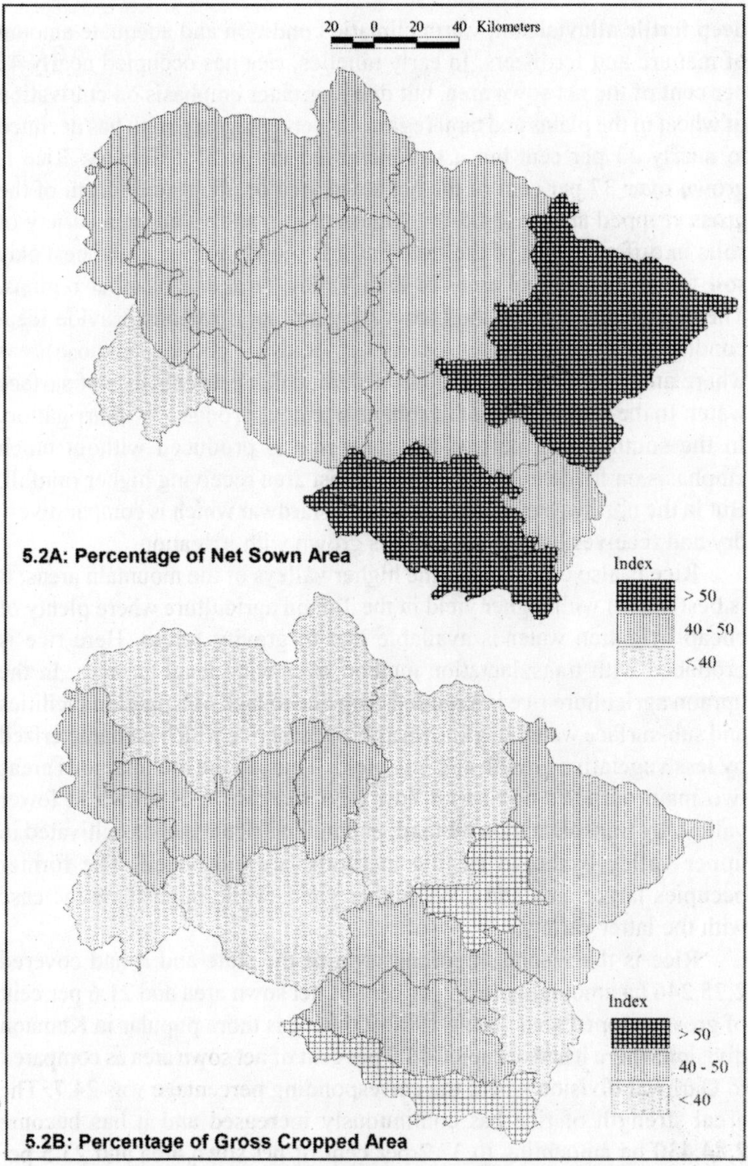

Fig. 5.2: Spatial Distribution of Wheat: 2005-06

deep fertile alluvial soil, warm climatic condition and adequate amount of manure and fertilizers. In early nineties, rice has occupied nearly 42 per cent of the net sown area, but due to greater emphasis on cultivation of wheat in the plains and tarai region, its percentage strength has declined to nearly 23 per cent but it has increased in the 21st Century. Rice is grown over 37 per cent of the net sown area or about one-fourth of the gross cropped area in 2005-06. Although it is cultivated on a variety of soils in different area of the state but it is mostly grown in the best clay soil found in the level areas of Dun Valley, tarai and bhanbar regions. The calcareous soils of the Dun Valley and tarai regions provide ideal condition for rice cultivation but higher yields are obtained in those areas where an impervious sub-stratum occurs which prevents loss of surface water. In the porous soils of the bhanbar, rice is produced with irrigation. In the southeastern part of the state rice is produced without much emphasis on irrigation because it is moist area receiving higher rainfall. But in the northwestern part specially in Hardwar which is comparatively dry and receives less rainfall, rice is grown with irrigation.

Rice is also cultivated in the higher valleys of the mountain areas. It is best grown with higher yield in the Talaon agriculture where plenty of cheap irrigation water is available due to gravity forces. Here rice is produced with transplantation method over wide level terraces. In the upraon agriculture rice is a rainfed crop due to lack of irrigation facilities and sub-surface water. It is known as broadcast rice which is characterized by less vegetative growth and low yield. Thus, in the mountainous areas two main varieties of rice i.e. low land rice which is grown in lower valleys by transplantation method, and upland rice which is cultivated in upper valley by broad casting methods, are cultivated. The former occupies larger area and gives higher yield, while opposite is the case with the latter variety.

Rice is the second important crop of the state and it had covered 2,75,246 ha amounting to 35 per cent of net sown area and 21.6 per cent of gross cropped area during 1992-93. Rice is more popular in Kumaon division where it had occupied 46.3 per cent of net sown area as compared to Garhwal division where the corresponding percentage was 24.7. The areal strength of rice has continuously increased and it has become 2,84,430 ha amounting to 37.2 per cent of net sown area and 23.5 per cent of gross cropped area during 2005-06. The percentage share of Kumaon has increased to 51 per cent, while it was 23.5 per cent in case of Garhwal division (Appendix-IX).

At district level its percentage strength to net sown area varies from as low as 6 per cent in Almora to as high as nearly 74 per cent in US Nagar (Table 5.3). The spatial distribution of rice cultivation in the state is represented in Fig.5.3A which shows that there are 4 districts viz. Rudraprayag (41%). Bageshwar (59%), Pithoragarh (49.4%) and US Nagar (73.8%) where it occupies more than 40 per cent of the net area sown in the districts.

In Rudraprayag the actual area under rice is lowest among all the districts of the state but as the larger percentage of net sown area is devoted to the cultivation of rice, its percentage has gone upto 41.00 per cent. Almost similar is the case in Pithoragarh and Bageshwar districts also. However, in all the three mountainous districts rice in mostly cultivated in the lower river valleys of Mandakini, Dhauliganga, Goriganga Ramganga and Saryu in Talaon system of cultivation. In lower areas of the upper part of the river valleys, rice is cultivated in Upraon system.

Table 5.3: Spatial Distribution of Rice : 2005-06

District/Division	*Area in ha*	*% to N.S.A.*	*% to G.C.A.*
1. Chamoli	10,219	31.39	22.49
2. Dehradun	12,344	25.49	16.75
3. Hardwar	17,137	14.21	10.20
4. Pauri Garhwal	22,335	27.41	17.91
5. Rudraprayag	8,016	41.07	28.63
6. Tehri Garhwal	12,164	20.58	13.37
7. Uttarkashi	10,211	33.13	21.50
Total: Garhwal Division	**92,426**	**23.56**	**15.98**
8. Almora	21,212	6.20	16.98
9. Bageshwar	15,699	59.09	33.00
10. Champawat	9,395	39.20	23.91
11. Nainital	12,378	25.93	15.14
12. Pithoragarh	22,696	49.44	28.37
13. US Nagar	1,10,624	73.82	42.46
Total: Kumaon Division	**1,92,004**	**51.20**	**30.28**
Total: Uttarakhand	**2,84,430**	**37.06**	**23.46**

Source: Compiled and Computed by the author.

The region of moderate concentration of rice cultivation with 25-40 per cent of net sown area, covers 6 districts including Champawat (39.2%), Uttarkashi (33.1%), Chamoli (31.4%), Pauri Garhwal (27.4%), Nainital (25.9%) and Dehradun (25.5%). Although all these districts are linked with one and other but extend from foothill plain to Trans-Himalayan zone.

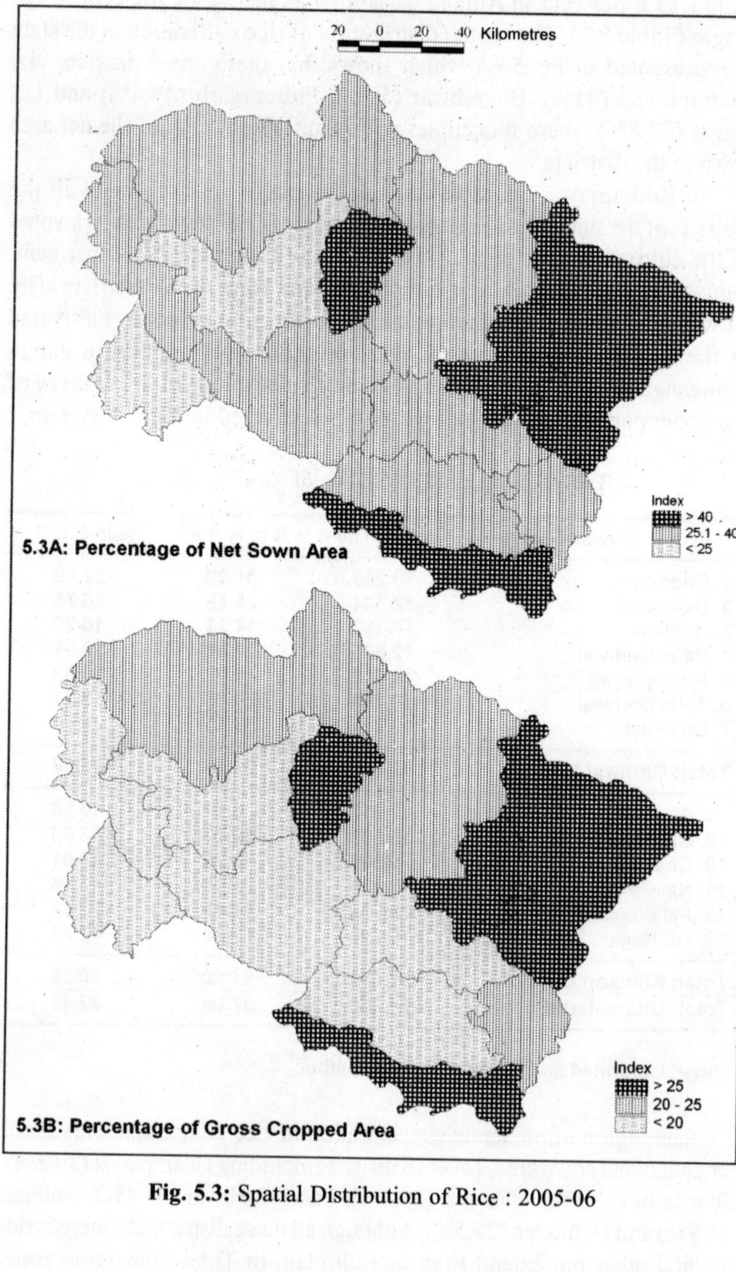

Fig. 5.3: Spatial Distribution of Rice : 2005-06

Here rice is grown in Talaon region of the river valleys. In Dehradun best-quality basmati rice is grown in Dun Valley and is mainly exported, while in Champawat it is grown in fertile lower valley of Saryu and Lodhia river system. In the region of low concentration of rice, there are only three districts viz. Hardwar (14.2%), Tehri Garhwal (20.6%) and Almora (16.2%) where cultivation of rice occupies less than 15 per cent of net sown area. In Tehri Garhwal it is the third important crop following wheat and Madua, and is mostly cultivated in the fertile lower parts of Bhagirathi and Bhillanganga, river system. In Hardwar district the lowest percentage under rice is because here larger area is given to the cultivation of sugarcane (42%) and wheat (25%) which are more productive and profitable commercial crops. The percentage occupancy of rice to gross cropped area is shown in Fig.5.3B which shows a very truncated and complex pattern of distribution. The ranks of the districts has also got changed.

It may be concluded from the above discussion that the larger area under rice cultivation is found in the plain, tarai and bhanbar areas but in some of the mountainous district the percentage strength of rice is higher because of lesser area under cultivation. US Nagar which produces export quality rice is the only district which devotes nearly 73 per cent of its net sown area to fine quality rice cultivation. The state has the potentiality to enhance the areal strength of rice with the application of modern agricultural technology. Therefore, its areal strength should be increased following appropriate modern agricultural techniques without adversely affecting the ecological balance of the region.

5.3.3 *Madua*

Madua—small millet is a coarse grain food crop of the inhabitants of the state. It requires lesser amount of rainfall and low temperature. It can be grown on a variety of poor quality soil. It can withstand more harder environmental conditions as compared to wheat and rice. Madua is a rainfed kharif crop grown with little care. Generally, it is cultivated in the areas which are unfit for fine crops of wheat and rice, and it gives higher yields also.

Before the induction of new technology in agriculture, Madua was widely grown in the state over a large area, and was a staple food but now its areal strength is declining. In 1992-93 Madua was grown over an area of 1,39,315 ha accounting to nearly 18 per cent of the net sown area or 11.0 per cent of gross cropped area. Out of this, more than 18 per cent was in Garhwal division and nearly 17 per cent of net sown area in Kumaon division. It was sown in Katil areas in September by clearing the land from forest in March. It used to give higher yield on origin land. But as it damages the vegetative cover and induces soil erosion, this practice is gradually abandoned. During 2003-04 its areal strength has declined to 1,26,423 ha amounting to 16.6 per cent of net sown area or nearly 10 per cent of the gross cropped area. Garhwal division has accounted nearly 18 per cent, while the share of Kumaon division was 15.0 per cent of net sown area. Almost the similar status of Madua has been noticed during 2005-06 also. It has become quite clear from this analysis that the cultivation of Madua is more popular and has occupied more area in Garhwal division as compared to Kumaon division (Appendix-IX).

The spatial distribution of Madua cultivation is highly uneven at district level and it ranges from as low as negligible in Hardwar to as high as nearly 47.6 per cent of net sown area in Almora (Table 5.4). Fig.5.4 representing the percentage strength of Madua at district level, shows that there are 2 contiguous districts viz. Pauri Garhwal (34.8%) and Almora (41.6) which have more than 30 per cent of net sown area under Madua cultivation. In all these districts Madua is cultivated in Upraon farming system in the upper parts of the river valleys. Here it is cultivated without irrigation. Its cultivation also extends in Katil system of farming with extremely poor yield.

Table 5.4: Spatial Distribution of Madua : 2005-06

District/Division	Area in ha	% to N.S.A.	% to G.C.A.
1. Chamoli	8,665	26.61	19.07
2. Dehradun	2,692	5.55	3.65
3. Hardwar	0.00	0.00	0.00
4. Pauri Garhwal	28,363	34.81	22.75
5. Rudraprayag	5,791	29.30	20.43
6. Tehri Garhwal	13,733	23.23	15.10
7. Uttarkashi	7,022	22.78	14.79
Total: Garhwal Division	**66,194**	**16.86**	**11.45**
8. Almora	33,648	41.57	26.93
9. Bageshwar	5,840	21.98	12.28
10. Champawat	6,655	22.77	16.94
11. Nainital	3,481	7.29	4.26
12. Pithoragarh	8,511	18.54	10.64
13. US Nagar	50	0.03	0.02
Total: Kumaon Division	**58,185**	**16.52**	**9.18**
Total: Uttarakhand	**1,24,379**	**16.40**	**10.26**

Source: Compiled and Computed by the author.

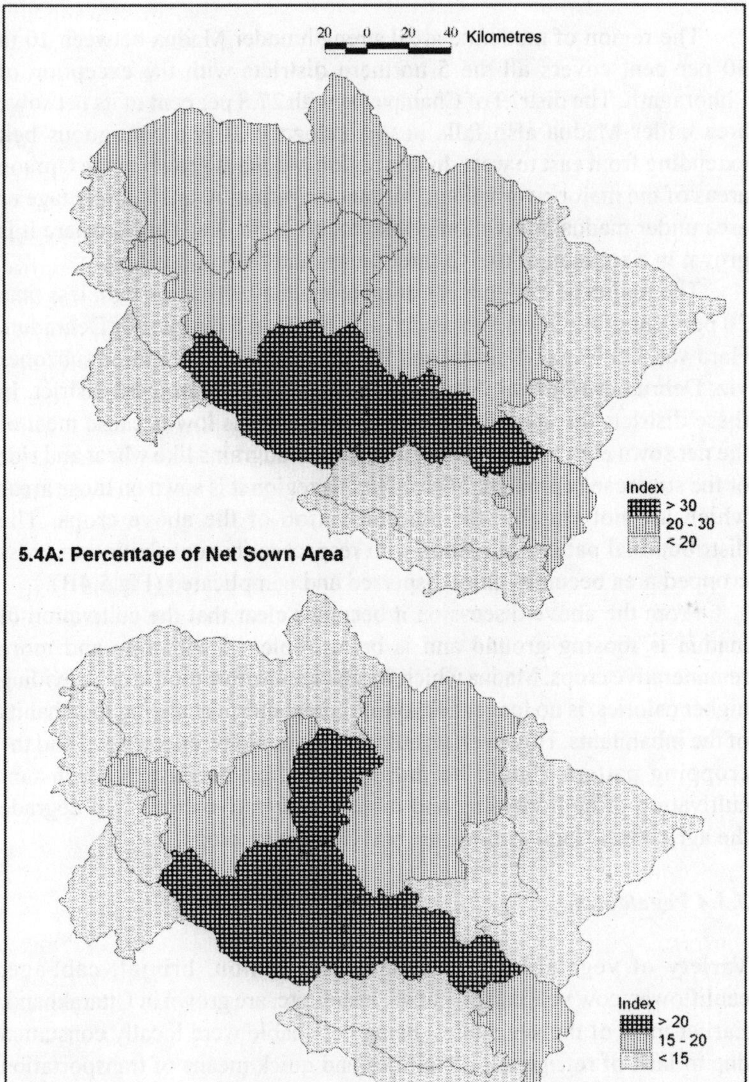

Fig. 5.4: Spatial Distribution of Madua : 2005-06

The region of moderate areal strength under Madua between 20 to 30 per cent covers all the 5 northern districts with the exception of Pithoragarh. The district of Champawat with 27.8 per cent of its net sown area under Madua also falls in this category. It is a contiguous belt extending from east to west. In this region Madua is grown in the Upraon areas of the major river valleys without irrigation. A small percentage of area under madua cultivation is also found in the Katil areas where it is grown in a rotation of two to three years, and the yield is low.

The region of low concentration of madua cultivation with less than 20 per cent of net sown area under it, is found in 5 districts of Dehradun, Hardwar, US Nagar, Nainital and Pithoragarh. There are three subzones viz. Dehradun-Hardwar, Nainital-US Nagar and Pithoragarh district. In these districts the area under madua cultivation is low because most of the net sown area is devoted to either fine foodgrains like wheat and rice or the sugarcane commercial crop. In this region it is sown on those areas which are not suitable for the cultivation of the above crops. The distributional pattern of madua with respect to its contribution to gross cropped area becomes more dispersed and complicated (Fig.5.4B).

From the above discussion it becomes clear that the cultivation of madua is loosing ground and is being replaced with fine and more remunerative crops. Madua which was more nutritive food crop providing higher calories, is no longer a favourite cereal due to changing food habits of the inhabitants. However, its cultivation should be encouraged and the cropping pattern should be more diversified because the constant cultivation of high yielding and more remunerative crops will degrade the agricultural land as these are soil exhaustive crops.

5.3.4 Vegetables

Variety of vegetables including potato, onion, brinjal, cabbage, cauliflower, cow pea, carrot, radish, beans, etc. are grown in Uttarakhand. Earlier most of the vegetables, being perishable were locally consumed due to lack of refrigeration facilities and quick means of transportation for taking it to markets of the down plains. Hence, lesser area was devoted to its cultivation. But now due to quick means of transport and refrigeration facilities, perishable vegetables could be easily transported to the market centres in the plain. Therefore, the area under vegetable cultivation has considerably increased in the recent past. Owing to the favourable climatic conditions for the cultivation of vegetables in the summer season, the state has an added advantage to produce vegetables when it is in great demand in the plains. Vegetables are mostly grown around the settlement sites because of high fertility of soil, ready market and easy care of the crops. It is also grown in the fertile plains, tarai and bhanbar areas with the help of irrigation, and in lower river valleys in mountainous areas.

There has been a lot of spatio-temporal variations in the areal strength of vegetables (Appendix-IX). In 1992-93 it had occupied 22,978 ha of land amounting to nearly 2.9 per cent of the net sown area in the state, while it had covered 11,788 ha in Garhwal and 11,190 ha in Kumaon division amounting to 2.9 and 3.0 per cent of the net sown area respectively in the same year. The areal strength has increased to 33,147 ha accounting to 4.35 per cent of net sown area or 2.7 per cent of gross cropped area in 2003-04. In Garhwal division the areal coverage of vegetables was 16,706 ha., while in Kumaon division it was 16,441 ha. contributing 4.20 and 4.56 per cent of net sown area and 2.8 and 2.7 per cent of gross cropped area respectively during the same year. During 2005-06 the area under vegetable has further increased to 34,802 ha amounting to 4.5 per cent of net sown area and 2.9 per cent of gross cropped area. The interesting fact is that the area under vegetables has been much more in hilly areas as compared to plains. During 2003-04 the area under vegetables in hill areas was 23,709 ha (71.5%) as against 9,438 ha (28.5%) in plains, while during 2005-06 the corresponding figures were 25,200 ha (72%) and 9,602 ha (28.5%) in hill and plain areas respectively.

Table 5.5: Spatial Distribution of Vegetables : 2005-06

District/Division	*Area in ha.*	*% to N.S.A.*	*% to G.C.A.*
1. Chamoli	2,201	6.76	4.84
2. Dehradun	6,562	13.53	8.91
3. Hardwar	10,184	8.45	6.06
4. Pauri Garhwal	1,449	1.78	1.16
5. Rudraprayag	293	1.50	1.05
6. Tehri Garhwal	2,367	4.00	2.60
7. Uttarkashi	4,340	14.08	9.14
Total: Garhwal Division	**27,409**	**6.98**	**7.74**
8. Almora	2,285	2.82	1.83
9. Bageshwar	719	2.71	1.51
10. Champawat	1,569	6.55	3.99
11. Nainital	12,325	5.82	15.08
12. Pithoragarh	2,686	5.83	3.36
13. US Nagar	3,705	2.47	1.42
Total: Kumaon Division	**23,289**	**6.21**	**3.67**
Total: Uttarakhand	**50,698**	**6.60**	**4.18**

Source: Compiled and Computed by the author.

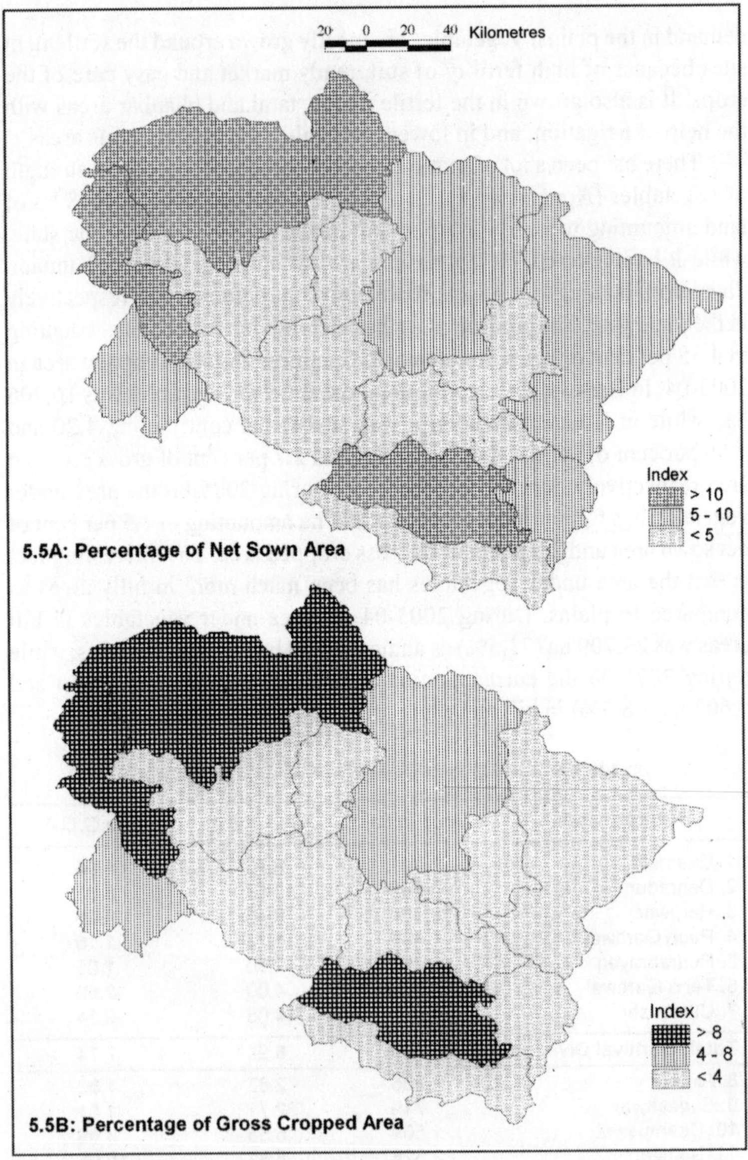

Fig. 5.5: Spatial Distribution of Vegetables : 2005-06

Spatial distribution of vegetables cultivation in Uttarakhand is given in Table 5.5 and is shown in Fig. 5.5 which shows a lot of inter-district variations in it. The highest concentration of vegetable cultivation is found in 3 districts viz. Uttarkashi (14%), Dehradun (13.5%) and Nainital (25.8%) where it covers more than 10 per cent of net sown area. In Nainital high percentage of urban population creating more demand for vegetables, proximity to the markets of the plain leading to better connectivity, fertile soil and conducive climatic conditions are the main causes for its higher areal strength, while in Uttarkashi fairly cold climate and lesser area under cultivation leading to increase in percentage contribution of vegetables cultivation are responsible for it. It is mostly grown around Bhatwari and Uttarkashi towns. In Nainital, the lower valley area around Bhowali and the plain of Haldwani tahsils are the major areas of vegetables cultivation. In Dehradun also the higher percentage under vegetables cultivation may be accounted to the local market of Dehradun town, its better connectivity with plain and favourable geographical conditions.

The region of moderate cultivation of vegetables ranging between 5 to 10 per cent of net sown area, covers 4 districts viz. Chamoli (6.7%), Hardwar (8.5%), Champawat (6.6%) and Pithoragarh (5.8%) (Fig. 3.3A). It constitutes two pockets being separated by Rudraprayag which may also be included in it with slight-modification in grouping the districts. However, one lies in northeastern part consisting of 3 districts, while the other is found in Hardwar district. There are varied reasons ranging from strategic location, conducive climatic conditions and fertile soil for higher percentage.

The region of low concentration with less than 5.0 per cent of net sown area covers 6 districts and is found in two clusters. The largest tract consisting of 5 districts extends from fertile alluvial plain of Garhwal to the central mountain ranges of Rudraprayag and Bageshwar. In this region also the higher concentration is found around the settlements and fertile lower valleys of the rivers. The share of vegetables in the gross cropped area in each district varies a lot and shows a dispersed and complicated distributional pattern (Fig. 5.5B).

From the above discussions it becomes clear that due to favourable environmental conditions, the state has huge potentiality for the development of vegetable farming provided the proper transport connectivity is developed and marketing facilities and infrastructure are improved properly.

5.3.5 Maize

Maize which is also known as Indian corn is one of the staple foods of the inhabitants of the state. It can be reaped in short-term and is, therefore, of special value as one of the first crop available for food. Maize is generally grown after the first shower of the monsoon. A long break in the rainfall or excessive fall and floods damage the crop. Waterlogging is equally harmful. Hence, the crop is grown on higher reaches of the river valleys with reasonable slope. Well drained and generally light soils are best suited for maize in preferences to clay or clayey loam. Maize is a rainfed crop grown without irrigation across the state. The cultivation of maize is somewhat laborious business. After the formation of cobs, it has to be protected from birds, stray animals and even from human beings. Maize cobs are generally picked up for being used partly as green cobs and partly as ripe cobs. The stalk of maize provides a nutritious fodder, if they are cut while green, but as the crops is grown for its grain, the stalk is used for dry fodder and fuel.

Table 5.6: Spatial Distribution of Maize : 2005-06

District/Division	Area in ha.	% to N.S.A.	% to G.C.A.
1. Chamoli	195	0.60	0.43
2. Dehradun	10,917	22.51	14.82
3. Hardwar	1,335	1.11	0.79
4. Pauri Garhwal	2,305	2.83	1.85
5. Rudraprayag	194	0.99	0.69
6. Tehri Garhwal	1,972	3.34	2.17
7. Uttarkashi	681	2.21	1.43
Total: Garhwal Division	**17,599**	**4.48**	**3.04**
8. Almora	1,887	2.33	1.51
9. Bageshwar	1,468	5.53	3.09
10. Champawat	853	3.56	2.17
11. Nainital	5,830	12.21	7.13
12. Pithoragarh	3,585	7.81	4.48
13. US Nagar	239	0.16	0.09
Total: Kumaon Division	**13,862**	**3.70**	**2.19**
Total: Uttarakhand	**31,461**	**4.10**	**2.60**

Source : Compiled and Computed by the author.

Maize is widely grown across the state in upland infertile land. There has been fluctuations in the areal strength of maize over the years. It had occupied 33,065 ha or nearly 5 per cent of the net area sown during 1992-93 in the state, although it was nearly 8 per cent and 3 per cent in Garhwal and Kumaon divisions respectively. During the year 2003-04

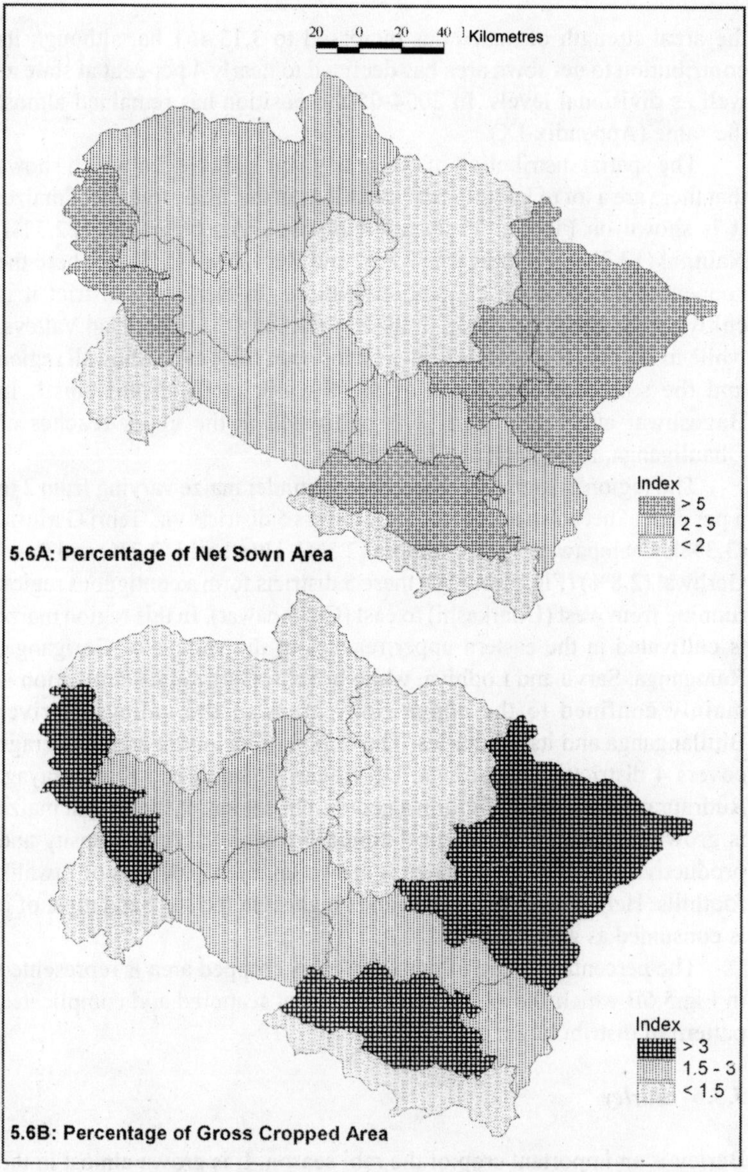

Fig. 5.6: Spatial Distribution of Maize : 2005-06

the areal strength of maize has increased to 3,13,461 ha, although its contribution to net sown area has declined to nearly 4 per cent at state as well as divisional levels. In 2004-05 the position has remained almost the same (Appendix-IX).

The spatial distribution of maize is given in Table 5.6 which shows that there are a lot of inter-district variations in the areal strength of maize. It is shown on Fig.5.6. There are 4 districts viz. Dehradun (22.5%), Nainital (12.2%), Pithoragarh (7.8%) and Bageshwar (5.5%), where the concentration of maize cultivation is more. In Dehradun district it is cultivated on the slopy flanks of the Siwaliks and the upper Dun Valleys, while in Nainital district it is grown on the upper slopes of Bhowali region and the southwestern slopes of the Siwalik in Haldwani tahsil. In Bageshwar and Pithoragarh it is cultivated in the upper reaches of Dhauliganga, Ramganga and Sarju rivers.

The region of moderate areal strength under maize varying from 2 to 5 per cent of net sown area in the state covers 5 districts viz. Tehri Garhwal (3.3%), Champawat (3.6%), Almora (2.3%), Uttarkashi (2.2%) and Pauri Garhwal (2.8%) (Fig. 5.6A). All these 5 districts form a contiguous region running from west (Uttarkashi) to east (Champawat). In this region maize is cultivated in the eastern upper reaches of the valleys of Goriganga, Ramganga, Saryu and Lodhiya, while in Tehri Garhwal its cultivation is mainly confined to the upper slopy areas of the valleys of river Bhillanganga and its tributaries. The region of its lowest areal coverage covers 4 districts viz. Hardwar, US Nagar, Chamoli and Rudraprayag. Rudraprayag and Chamoli form a contiguous region. It shows that maize is grown in variety of ecological conditions with varying intensity and productivity. In US Nagar maize is grown on infertile slopes of Siwalik foothills. Here, the yield of the crop is highest in the state and most of it is consumed as green cobs.

The percentage share of maize to gross cropped area is represented in Fig.5.6B which shows a little different but scattered and complicated pattern of distribution.

5.3.6 Barley

Barley is an important crop of the rabi season. It is grown almost in the same climatic conditions as wheat but it is more adaptable and has a wider range of cultivation than wheat. Moderate temperature with well distributed rainfall and plenty of sunshine during ripening period are suitable climatic conditions. It is a shallow rooted crop and it grows well on light shallow infertile soil. It is a relatively shorter growing season crop and is preferred over wheat in many areas specially in the upland cultivation. It is also cultivated in high altitude single crop zone where in many areas it is produced with the help of irrigation. In upraon it is generally sown side by side with wheat but it is also cultivated at higher altitude than wheat.

Table 5.7: Spatial Distribution of Barley: 2005-06

District/Division	*Area in ha*	*% to N.S.A.*	*% to G.C.A.*
1. Chamoli	1,547	4.75	3.40
2. Dehradun	1,052	2.17	1.43
3. Hardwar	23	0.02	0.01
4. Pauri Garhwal	6,777	8.32	5.44
5. Rudraprayag	1,056	5.41	3.77
6. Tehri Garhwal	2,297	3.89	2.53
7. Uttarkashi	305	0.99	0.64
Total: Garhwal Division	**13,057**	**3.33**	**2.26**
8. Almora	2,837	3.50	2.27
9. Bageshwar	1,786	6.72	3.75
10. Champawat	1,393	5.81	3.54
11. Nainital	940	1.97	1.15
12. Pithoragarh	3,382	7.37	4.23
13. US Nagar	25	0.02	0.01
Total: Kumaon Division	**10,363**	**2.76**	**1.63**
Total: Uttarakhand	**23,420**	**3.05**	**1.93**

Source: Compiled and Computed by the author.

Barley is widely grown crop cultivated in all the districts of the state with a lot of spatio-temporal variations in its areal strength. In 1992-93 it was grown over 27,770 ha or more than 4 per cent of the net area sown in the state, while it had covered 14,622 ha and 13,148 ha accounting to 5 and 3.8 per cent in Garhwal and Kumaon divisions respectively. During 2003-04 its areal strength has declined due to more emphasis being given to the cultivation of fine grain crops like wheat and it was almost 3 per cent of the net sown area at state and divisional levels, while its percentage strength to gross cropped area was around 2.5 per cent at all the two levels i.e. state and division. During 2005-06 the areal strength of barley has remained the same at division and at state levels (Appendix-IX).

Table 5.7 gives the areal strength of barley at district level and it is represented in Fig.5.6 which shows that there are a lot of inter-district variations in it during 2005-06. These are 3 districts viz. Pauri Garhwal

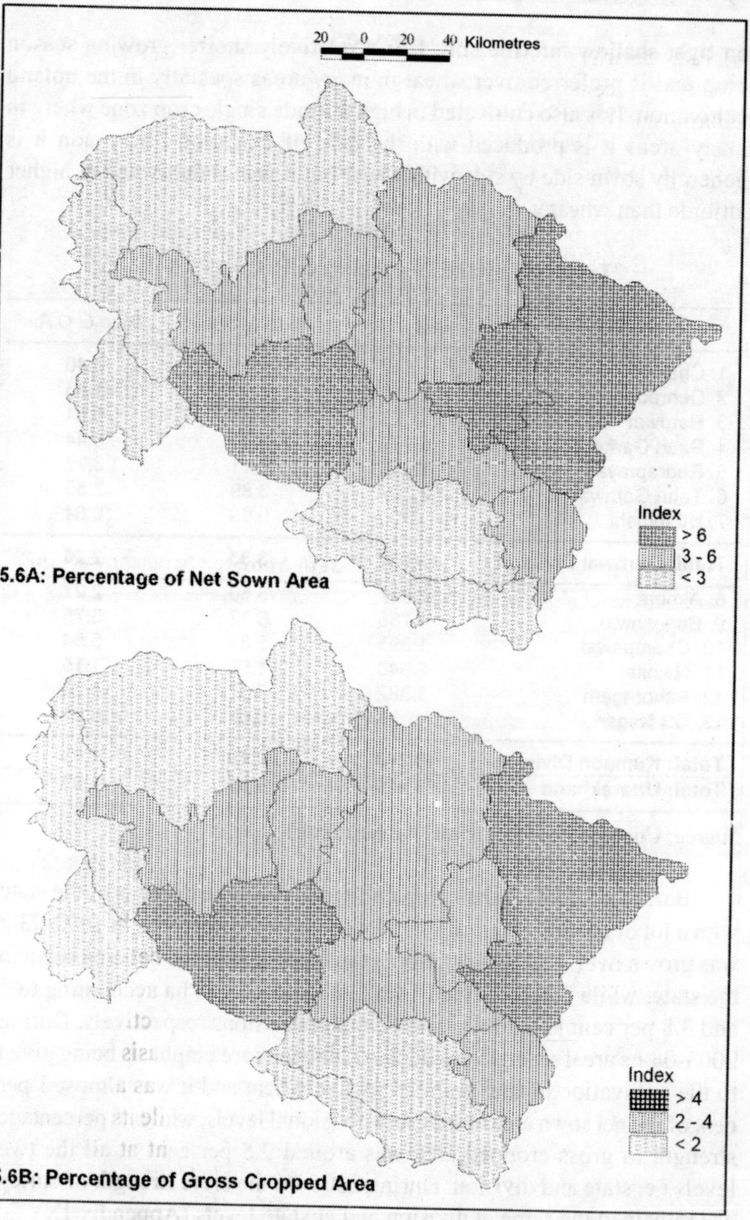

Fig. 5.7: Spatial Distribution of Barley: 2005-06

(8.3%), Pithoragarh (7.3%) and Bageshwar (6.7%) where barley occupies more than 6 per cent of the net sown area, while only in 2 districts of Pauri Garhwal (5.5%) and Pithoragarh (4.2%) it has covered more than 4 per cent of gross cropped area. In Pauri Garhwal district barley is grown in the tarai area, while in Bageshwar and Pithoragarh it is grown in the Dhauliganga and Goriganga valleys.

The region of moderate areal strength covering 5 district in a contiguous belt, extends from Tehri Garhwal in the west to Champawat in the east. It is mostly cultivated in the Upraon area of the river valleys in this region. The region of low concentration of barley cultivation extends over 5 districts forming three clusters. Nainital and US Nagar where it is grown in tarai and foothill plains form one cluster, while Dehradun and Hardwar constitute the other one. Here barley is grown in tarai and Dun Valley which are not very conducive for wheat cultivation. The percentage strength of barley to gross cropped area is represented in Fig.5.5B which shows almost identical pattern to net sown area.

During the last three decades the areal strength of barley has shown a steadily declining trend due to emphasis being given to the cultivation of wheat and other more fine and remunerative crops. However, the cultivation of barley should be encouraged in crop rotation system because it is a less soil exhaustive crop.

5.3.7 Pulses

Variety of pulses are grown in Uttarakhand but the major ones are gram, arhar, tur, peas, etc. Pulses, besides being more nutritive providing sufficient amount of protein, also enrich the soil. Pulses cover 51,560 ha of land amounting to nearly 7 per cent of net area sown and 4.3 per cent of gross cropped area in the state and almost similar is the share of pulses at divisional level. There has also been almost consistency in the areal strength of pulses over a period of more than a decade e.g. it was 52,502 ha in 1992-93 and 51,560 ha in 2005-06. But there has been spatio-temporal variations in its areal strength and its share in the net area sown at district level. During the agricultural year 2005-06 pulses had occupied as high as nearly 13.5 per cent of net area sown and 9 per cent of gross cropped area in US Nagar, while it was only 1.3 per cent and 0.93 per cent in Hardwar district respectively during the same time period (Appendix-IX).

The spatial distribution of percentage strength of pulses over net sown area and gross cropped area during 2005-06 is represented in Fig.5.8

Table 5.8: Spatial Distribution of Pulses : 2005-06

District/Division	*Area in ha.*	*% to N.S.A.*	*% to G.C.A.*
1. Chamoli	1,635	0.21	0.13
2. Dehradun	3,522	0.46	0.29
3. Hardwar	1,564	0.20	0.13
4. Pauri Garhwal	7,259	0.95	0.60
5. Rudraprayag	683	0.09	0.06
6. Tehri Garhwal	7,783	1.01	0.64
7. Uttarkashi	4,158	0.54	0.34
Total: Garhwal Division	**26,604**	**3.47**	**2.19**
8. Almora	5,597	0.73	0.46
9. Bageshwar	2,258	0.29	0.19
10. Champawat	3,208	0.42	0.26
11. Nainital	2,969	0.39	0.24
12. Pithoragarh	7,289	0.95	0.60
13. US Nagar	3,635	0.47	0.30
Total: Kumaon Division	**24,956**	**3.25**	**2.06**
Total: Uttarakhand	**51,560**	**6.72**	**4.25**

Source: Compiled and Computed by the author.

which shows that there are 4 districts viz. Uttarkashi, Tehri Garhwal, Pithoragarh and Champawat where pulses have shared more than 10 per cent of the net sown area and more than 8 per cent of the gross cropped area. There are two such pockets consisting of two districts each lying in north western and eastern part of the state. The zone of moderate concentration of pulses cultivation including 6 districts of Dehradun, Tehri Garhwal, Nainital, Almora, Bageshwar and Chamoli form a contiguous region and spreads from Trans-Himalaya in the north to foothill plains in the south of the state because of variety of pulses requiring different environment are cultivated in the state (Fig.5.8B).

The distributional pattern of moderate zone of pulses cultivation in case of gross cropped area is somewhat different. It covers four districts of Dehradun, Garhwal, Almora and Bageshwar lying in a contiguous belt-running from west to east in the southern part of the state. The zone of low concentration of pulses with less than 5 per cent are the districts of Hardwar, Rudraprayag and US Nagar. It is widely scattered in case of its share in net sown area, while in case of gross cropped area it spreads over 5 districts and forms 3 subzones (Fig.5.8B). In view of the growing demand of protein rich food and deteriorating soil fertility, the cultivation of various varieties of pulses on more agricultural land depending upon the local environment should be encouraged.

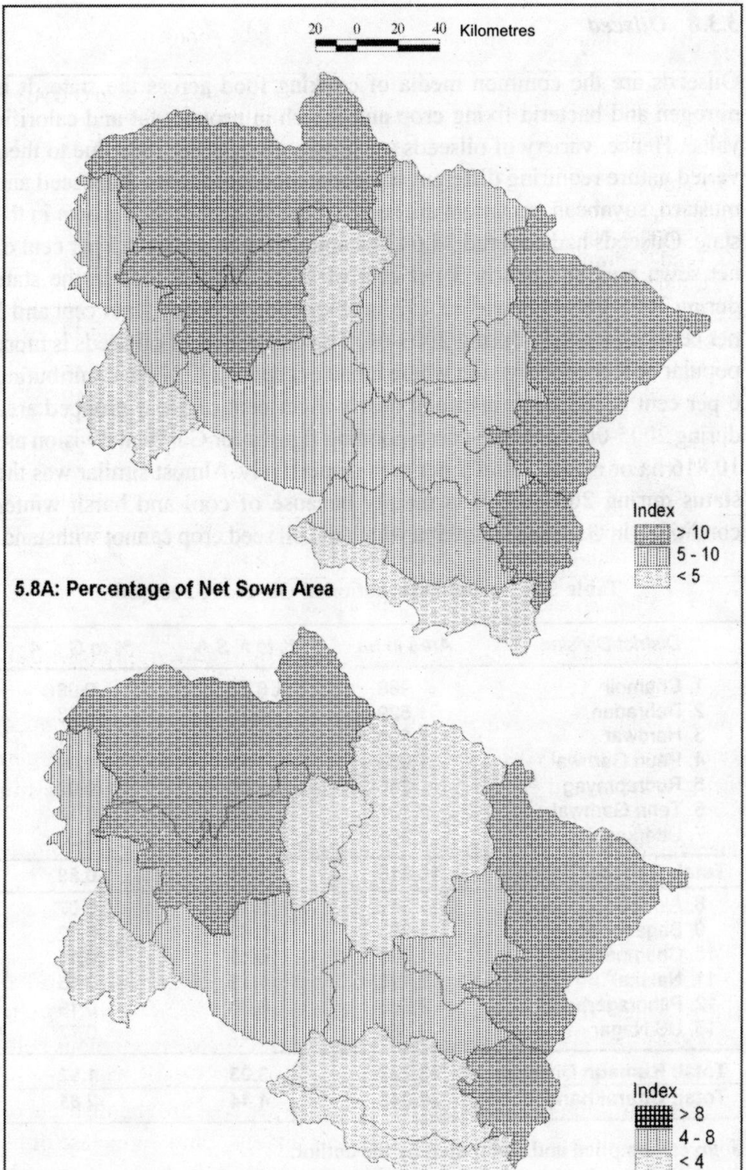

Fig. 5.8: Spatial Distribution of Pulses : 2005-06

5.3.8 Oilseed

Oilseeds are the common media of cooking food across the state. It is nitrogen and bacteria fixing crop and is rich in protein, fat and calorific value. Hence, variety of oilseeds are grown all over the state due to their varied nature requiring different environmental conditions. Rapeseed and mustard, soyabean, sesamum are some of the major oilseeds grown in the state. Oilseeds had covered 34,043 ha amounting to nearly 4.5 per cent of net sown area and nearly 3 per cent of gross cropped area in the state during 2005-06 which was 37,152 ha amounting to nearly 5 per cent and 3 per cent respectively during 2003-04. The cultivation of oil seeds is more popular in Kumaon division where it had occupied 23227 ha. contributing 6 per cent of net sown area and nearly 4 per cent of gross cropped area during 2005-06, while the corresponding figures for Garhwal division are 10,816 ha or nearly 3 and 2 per cent respectively. Almost similar was the status during 2003-04. It is mainly because of cool and harsh winter conditions in Garhwal Himalaya which the oil seed crop cannot withstand.

Table 5.9: Spatial Distribution of Oilseeds : 2005-06

District/Division	*Area in ha.*	*% to N.S.A.*	*% to G.C.A.*
1. Chamoli	986	0.13	0.08
2. Dehradun	1,529	0.20	0.13
3. Hardwar	2,500	0.33	0.21
4. Pauri Garhwal	1,033	0.13	0.09
5. Rudraprayag	266	0.03	0.02
6. Tehri Garhwal	2,137	0.28	0.18
7. Uttarkashi	2,365	0.31	0.20
Total: Garhwal Division	**10,816**	**1.41**	**0.89**
8. Almora	842	0.11	0.07
9. Bageshwar	203	0.03	0.02
10. Champawat	1,330	0.18	0.11
11. Nainital	9,125	1.19	0.75
12. Pithoragarh	2,288	0.30	0.19
13. US Nagar	9,389	1.22	0.77
Total: Kumaon Division	**23,227**	**3.03**	**1.92**
Total: Uttarakhand	**34,043**	**4.44**	**2.81**

Source: Compiled and Computed by the author.

Like all other crops, there are a lot of spatio-temporal variations in the areal strength of oilseeds at district level (Table 5.9). It has the highest percentage of net sown area (7.8%) and gross cropped area (4.9%) in US Nagar district, while the lowest percentage of 1.27 and 0.83 had been

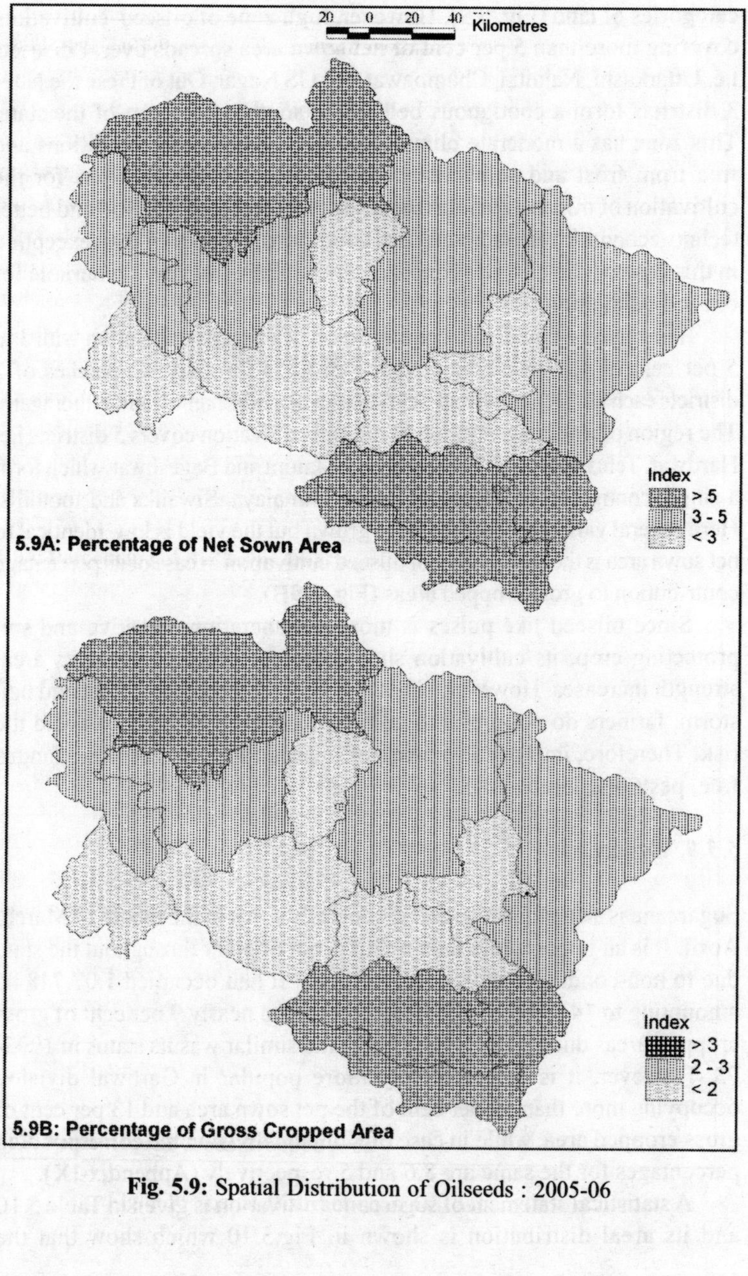

Fig. 5.9: Spatial Distribution of Oilseeds : 2005-06

recorded percentage in Pauri Garhwal respectively in case of both categories of land (Fig. 5.9). However, high zone of oilseed cultivation, covering more than 5 per cent of net sown area spreads over 4 districts i.e. Uttarkashi, Nainital, Champawat and US Nagar. Out of these the latter 3 districts form a contiguous belt in the southeastern part of the state. This zone has a moderate climate, comparatively warm conditions and free from frost and snowfall which are conducive conditions for the cultivation of oilseed. Besides these, this region has fertile soil and better techno-economic and socio-cultural conditions. Uttarkashi is an exception in this case where climatic condition and soil are not very favourable for oilseed cultivation.

The region of moderate concentration of oilseeds cultivation with 3 to 5 per cent of net sown area covers 4 districts forming two patches of 2 districts each i.e. Dehradun and Tehri Garhwal, and Chamoli and Pithoragarh. The region of low concentration of oilseed cultivation covers 5 districts i.e. Hardwar, Tehri, Garhwal, Rudraprayag, Almora and Bageshwar which form a single contiguous belt lying in Lesser Himalaya, Siwaliks and foothills. Here several varieties of oilseeds are grown but the yield is low. Identical to net sown area is the distribution of oilseed cultivation in case of its percentage contribution to gross cropped areas (Fig. 5.9B).

Since oilseed like pulses is more remuneration, nutritive and soil protecting crop, its cultivation should be encouraged so that its areal strength increases. However, since it is very susceptible to frost and hail storm, farmers do not prefer to cultivate this crop in order to avoid the risk. Therefore, improved varieties of oilseeds such as frost free, fungus free, pests and insects free.

5.3.9 Sugarcane

Sugarcane is an annual crop and is generally sown in the month of March/ April. It is an important cash crop but is not popular throughout the state due to non-conducive physical conditions. It had occupied 1,07,748 ha amounting to 14 per cent of net area sown and nearly 9 per cent of gross cropped areas during 2005-06. Almost the similar was its status in 1993-93. However, it is comparatively more popular in Garhwal division occupying more than 19 per cent of the net sown area and 13 per cent of gross cropped area, while in case of Kumaon division the corresponding percentages for the same are 8.6 and 5 respectively (Appendix-IX).

A statistical statement of sugarcane cultivation is given in Table 5.10 and its areal distribution is shown in Fig.5.10 which show that the cultivation of sugarcane is done only in 6 districts of tarai, bhanbar and foothills. In the remaining 7 districts it is not cultivated due to adverse climatic condition, surface configuration and poor thin soil. Table 5.10 shows a lot of variations in the areal strength of sugarcane. Its highest concentration is in Hardwar district which accounts nearly 66 per cent of the total area under sugarcane in the whole state because of its fertile alluvial soil, plenty of irrigation water and high level of agricultural technology (Fig. 5.10).

Table 5.10: Spatial Distribution of Sugarcane : 2005-06

District/Division	*Area Under Maize*	*% to N.S.A.*	*% to G.C.A.*
1. Chamoli	0	0.00	0.00
2. Dehradun	4,899	10.10	6.65
3. Hardwar	70,591	58.54	42.00
4. Pauri Garhwal	10	0.01	0.01
5. Rudraprayag	0	0.00	0.00
6. Tehri Garhwal	0	0.00	0.00
7. Uttarkashi	0	0.00	0.00
Total: Garhwal Division	**75,500**	**19.23**	**13.06**
8. Almora	0	0.00	0.00
9. Bageshwar	0	0.00	0.00
10. Champawat	7	0.03	0.02
11. Nainital	4,550	9.53	5.57
12. Pithoragarh	0	0.00	0.00
13. US Nagar	27,691	18.48	10.63
Total: Kumaon Division	**32,248**	**8.60**	**5.09**
Total: Uttarakhand	**1,07,748**	**14.04**	**8.89**

Source: Compiled and Computed by the author.

In Hardwar, nearly 59 per cent of the net area sown is devoted to sugarcane cultivation which covers 42 per cent of gross cropped area. It is followed by the district of US Nagar where 27,691 ha amounting to 18.5 per cent of net sown area or 10.6 per cent of gross cropped area are devoted to sugarcane cultivation. It has almost similar favourale conditions but wheat and rice are more common than sugarcane. The third district in rank of sugarcane cultivation is Dehradun where 4,899 ha area covering 10 per cent of net sown area or 6.7 per cent of gross cropped area is devoted to its cultivation. Here, the fertile calcareous soil is more suitable for sugarcane cultivation. It is grown in Vikasnagar and Rishikesh tahsils of Nainital where 4,550 ha land amounting to 9.5 per cent of net sown

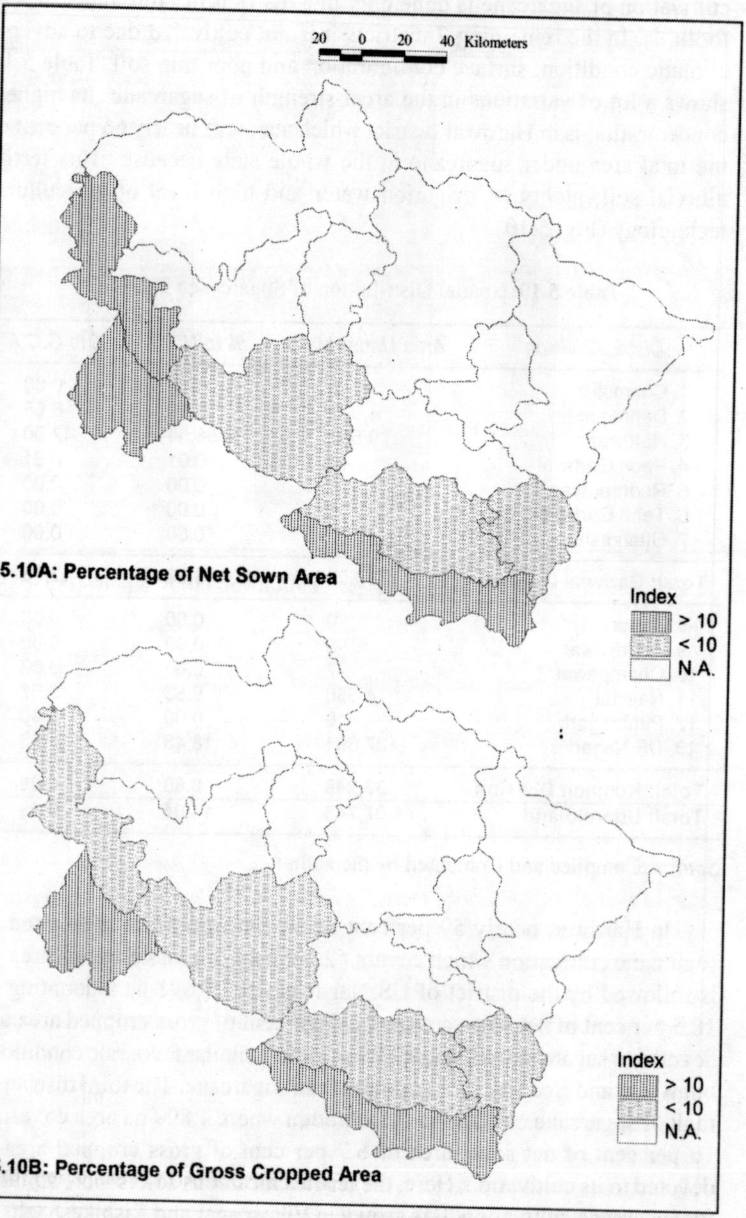

Fig. 5.10: Spatial Distribution of Sugarcane : 2005-06

area 5.6 per cent of gross cropped area is devoted to the cultivation of sugarcane. Here the cultivation of sugarcane is mainly found in tarai and foothill plains of Haldwani tahsils. Sugarcane is more remunerative but soil exhaustive crop also. It remain in the field all the year round. Hence, its cultivation in the state should not be encouraged.

5.4 Intensity of Cropping

Intensity of cropping is a measure to express the use of land during one agricultural year. It is mainly influenced by ecological, techno-economic and socio-cultural factors. By and large, the land is put to more intensive use in the areas where the agricultural land is scarce and its demand is more. In such areas land is put to maximum use at the given level of technology. This system gradually deteriorates the soil and the yield becomes low. It is, by and large, practiced in the developing countries with high density of population. On the contrary, almost opposite is the case where land is in abundance and its demand is low due to sparse density of population and large population engaged in secondary and tertiary activities. In this system the yield is high and the level of soil deterioration is low. This system is more prevalent in the developed countries with lower density of population (Chand, 1997).

Since the crops are being produced on the net sown area available, the intensity of cropping is measured with reference to it. There are many ways of measuring the intensity of cropping but the most common method of its measurement is given below:

$$\text{I.C.} = \frac{\text{GCL}}{\text{NSA}} \times 100$$

Where

I.C. = Intensity of Cropping
GCA = Gross Cropped Area
NSA = Net Sown Area

The state of Uttarakhand is a mountainous state having only 13 per cent of its total area as net sown area which is mostly found in the tarai, bhanbar, plains and the lower areas of the major river valleys. These areas, by and large are regions of high density of population. Agriculture is the basic economic activity of people in the absence of industrial units and other secondary and tertiary activities. The state also lacks in mineral resources. So the people have to earn their living from this limited land resource. Therefore, the intensity of cropping is, by and large, high in the state.

Table 5.11: Intensity of Cropping: 2005-06

District/Division	*Intensity of Cropping*		
	1992-93	*2003-04*	*2005-06*
1. Chamoli	153.67	148.11	139.56
2. Dehradun	162.40	155.98	151.89
3. Hardwar	149.84	144.35	139.38
4. Pauri Garhwal	155.38	154.09	152.99
5. Rudraprayag	150.27	160.33	143.47
6. Tehri Garhwal	159.31	157.96	153.83
7. Uttarkashi	150.27	155.10	154.10
Total: Garhwal Division	**150.00**	**151.84**	**147.30**
8. Almora	167.28	160.14	154.34
9. Bageshwar	167.28	180.10	179.08
10. Champawat	176.25	165/73	163.97
11. Nainital	170.25	168.70	171.24
12. Pithoragarh	176.45	184.41	174.27
13. US Nagar	170.25	171.25	173.85
Total: Kumaon Division	**168.10**	**170.13**	**169.10**
Total: Uttarakhand	**158.00**	**160.57**	**157.95**

Source: Compiled and Computed by the author.

Table 5.11 gives the intensity of cropping at district level and Fig.5.11 represents it. A cursory glance at the table and map shows a lot of spatio-temporal variations in the intensity of cropping. In 1992-93 it was nearly 158 per cent in the state, while in Garhwal and Kumaon divisions the figures are 150 and 168 per cent respectively. There was 8 districts with cropping intensity of more than state average, while in 5 districts it is less than it. During 2005-06 the cropping intensity was nearly 158 per cent in the state, while it was 147 and 169 per cent in Garhwal and Kumaon divisions. By and large, the cropping intensity is higher in Kumaon division as compared to Garhwal division because it supports a larger population on lesser net sown area and the level of technology is also higher. However, the high intensity of cropping zone with more the 165 per cent occurs in the districts of Pithoragarh, Bageshwar, Nainital and US Nagar which form two contiguous regions of 2 districts each. Nainital and US Nagar form one region where the intensity of cropping is high because of fertile alluvial soil capable of producing more than one crop

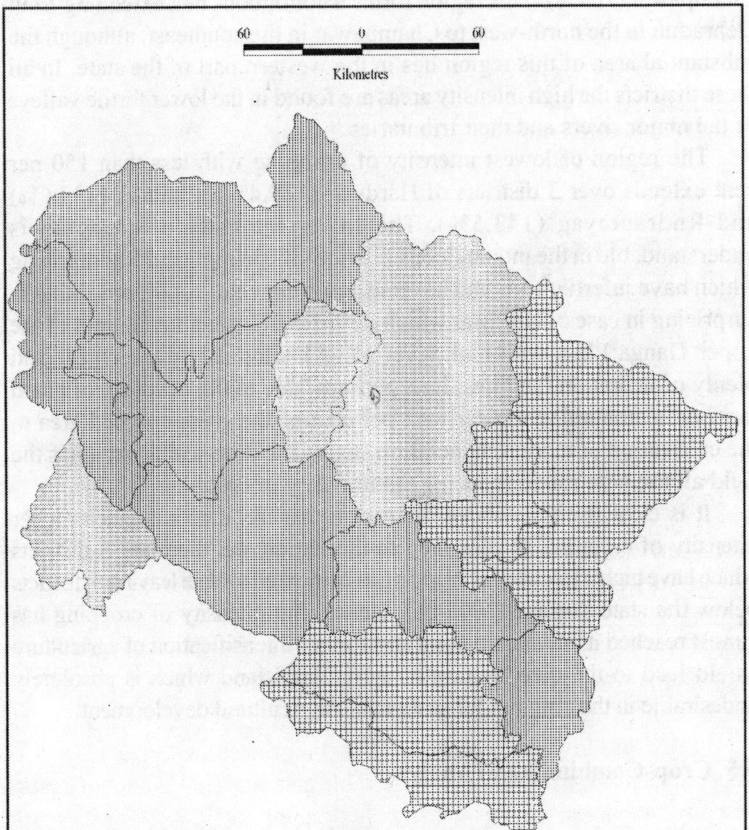

Fig. 5.11: Intensity of Cropping : 2005-06

in a year, high level of agricultural technology with sufficient water resources and enterprising farmers, while in the other region consisting of Pithoragarh and Bageshwar the high intensity of cropping is, by and large, due to the presence of fertile lower valleys of Dhauliganga, Goriganga, Ramganga and Sarju. The region of moderate intensity of cropping covers 6 districts of Dehradun (151.9%), Pauri Garhwal (153%), Tehri Garhwal (153%), Uttarkashi (154%), Almora (154.3%) and

Champawat (164%). This region forms a contiguous belt extending from Dehradun in the north-west to Champawat in the southeast, although the substantial area of this region lies in the western part of the state. In all these districts the high intensity areas are found in the lower fertile valleys of the major rivers and their tributaries.

The region of lowest intensity of cropping with less than 150 per cent extends over 3 districts of Hardwar (139.4%), Chamoli (139.6%) and Rudraprayag (143.5%). The lower intensity of cropping is understandable in the mountainous districts of Chamoli and Rudraprayag which have infertile soils and adverse environmental conditions, but it is surprising in case of Hardwar which lies in the fertile alluvial tract of the upper Ganga Plain with high level of technological advancement and plenty of irrigation facilities. It is, perhaps, due to the devotion of much large area amounting to nearly 42 per cent of the gross cropped area to the cultivation of more remuneration sugarcane crop which occupies the field all the year round reducing the intensity of cropping.

It is clear from the above discussion that the state has much higher intensity of cropping as compared to the nation and there are 5 districts which have higher intensity of cropping as compared to state leaving 8 districts below the state average (2005-06). Hence, the intensity of cropping has almost reached its saturation point and further intensification of agriculture would lead to the deterioration of agricultural land which is absolutely undesirable in the long run for sustainable agricultural development.

5.5 Crop-Combinational Analysis

Agriculture is, by and far, an important single use of land in most of the countries of the world, and the description and analysis of various problems relating to agricultural activities have contributed a great deal to the content of the regional and economic geography of these countries. Such studies have, however, confined themselves to the description of dominant individual crops rather than complex agricultural system, with the result that they usually suffer from overgeneralisations because the individual crops, except in relatively rare circumstances of extreme mono-culture, are not grown alone but characteristically they appear in combinations (Weaver, 1954). Moreover, it is significant to note that any process of development has an inherited property of pushing the economic system from simpler to more complex. This complexity is manifested by diversification of places in terms of their economic activities which can be measured by the different combinations of these activities.

Prior to the evolution of statistical technique, crop-combinations were identified on the basis of the percentage strength of the crops. Monoculture were identified on the basis of first ranking crops, 2 crop combinations on the basis of the first and second ranking crops, 3 crop combinations on the basis of first, second and third ranking crops and so on. But this intuitive classification based on subjective judgments like wheat zones, rice zones, industrial town, commercial town, etc., have no statistical basis. All the places, in fact, have multiple functions, but all of them may not be equally important at every place. Some of them may be very significant, while others might be quite inconsequential. A statistical basis is, therefore, necessary for identifying the group of significant functions, which represent the functional personality of an area (Mahmood, 1977).

Recently attention has been focused on combinational analysis of different elements in geographical studies which gives a scientific treatment to the geographical problems. Geographers are particularly interested in classifying the places into different combinations of dominant activities including those which pertain to agricultural production and landuse. This method ultimately minimizes the chances of oversimplified generalizations too (Mohammad, 1975 and Athawale, 1996).

A general understanding of the particular combination of crops and the relative importance of each in an area can be very helpful in interpreting some aspects of social and economic geography of the region. J.C. Weaver has pointed out three different lines in which crop-combinational studies may have more significance. Firstly, they are essential to an adequate understanding of individual crop geography; secondly, crop-combination in itself is an integrated reality that demands definitions and distributional analysis; and thirdly, such regions are essential for the construction of the still more complex structure of valid agricultural regions (Weaver, 1954).

Many methods have so far been devised by eminent scholars to identify and demarcate crop-combination regions such as Quartile Method (Johnson, 1958). Mean Positive Deviation Method (Pownall, 1953), Standard Deviation Method (Nelson, 1955), Minimum Deviation Method (Weaver, 1954), Maximum Positive Deviation Method (Rafiullah, 1965), Modified Minimum Deviation Method (Doi, 1959), Successive Quotients Method (Kostrowicki, 1972), Maximum Distance Method (Ayyar, 1969) and so on. In this study Doi's Modified Minimum Deviation Method has been used because several crop combinational studies (Mohammad, 1975; Siddiqui, 1975; Majeed, 1978) have found that in this method results are more realistic in comparison to other methods and they are not mere generalization. Moreover, the working with this method is much simple and easy.

5.5.1 The Ranking of Crops

Before discussing the crop-combination regions, it would be worthwhile to understand the ranking of crops which will facilitate the better understanding of crop strength. The relative position of crop with reference to its percentage contribution to the total cropped area has been ascertained for each district and the crops ranking first, second and third have been identified and mapped in Figure 5.12 for the year 2005-06 and a brief account of each is given below :

First Ranking Crops

It is clear from Fig.5.12 that wheat which occupies the highest percentage of gross cropped area in the state has ranked as first crop in as many as 11 districts (84.6%) leaving two districts of Hardwar and US Nagar. It shows that the entire state has conducive condition for cultivation of wheat. In the district of Hardwar lying in the Ganga plain, sugarcane has emerged as first ranking crop. Here, also wheat could have ranked first because of favourable conditions, but sugarcane is preferred due to its more remunerative value. In US Nagar rice has occupied the first rank due to favourable physical conditions, plenty of water resources and high level of technology. Moreover, it is very remunerative crop because high quality export variety of basmati rice is cultivated here (Fig.5.12A).

Second Ranking Crops

The picture becomes more diverse and complicated in case of second ranking crops. Rice, madua and wheat have emerged as second ranking crops. Rice emerges as second crops in 8 districts (61.5%). Out of these 5 districts viz. Uttarkashi, Rudraprayag, Chamoli, Bageshwar and Pithoragarh lie in the northern part of the state, while Nainital and Champawat lie in the southeast. All these seven districts form a contiguous belt. Dehradun is the only isolated district where fine quality expert variety of rice is grown due to favourable geographical conditions specially the quality of soil. Madua is the next crop which has ranked second in three districts viz. Tehri Garhwal, Pauri Garhwal and Almora which form a

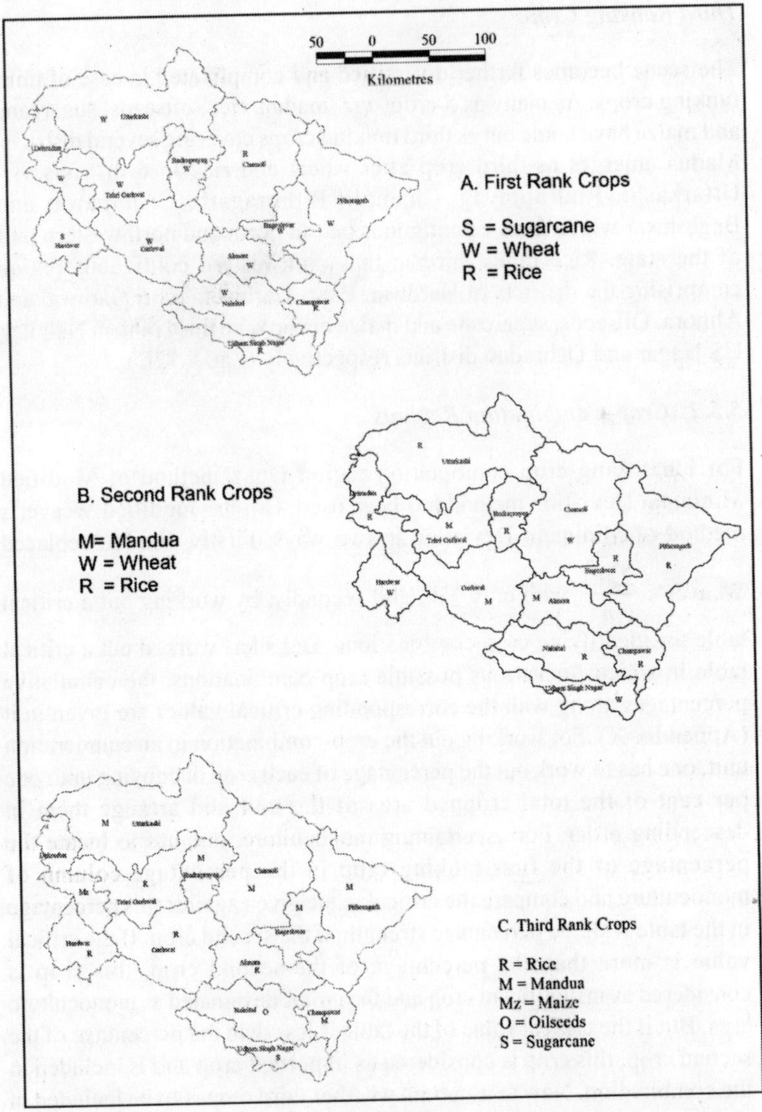

Fig. 5.12: Ranking of Crop : 2005-06

regular belt running from east to west in Siwalik Himalaya. Wheat has emerged as second important crop in Hardwar and US Nagar which have fertile alluvial soil and better techno-economic conditions (Fig. 5.12B).

Third Ranking Crops

The scene becomes further diversified and complicated in case of third ranking crops. As many as 5 crops viz. madua, rice, oilseeds, sugarcane and maize have come out as third ranking crops covering several districts. Madua emerges as third crop after wheat and rice in 6 districts viz. Uttarkashi, Rudraprayag, Chamoli, Pithoragarh, Champawat and Bageshwar which form a contiguous belt in north and northwestern part of the state. Rice ranks third in the southwestern contiguous region comprising the districts of Hardwar, Pauri Garhwal, Tehri Garhwal and Almora. Oilseeds, sugarcane and maize occupy the third rank in Nainital, US Nagar and Dehradun districts respectively (Fig. 5.12C).

5.5.2 *Crop-Combination Regions*

For identifying crop combination region Doi's method of Modified Minimum Deviation method has been used. Doi has modified Weaver's method of Minimum Deviation in two ways. Firstly, she has replaced Weaver's $\frac{\sum d^2}{n}$ with only $\sum d^2$, and secondly, by working out a critical table for identifying crop-combinations. Doi's has worked out a critical table in which for various possible crop combinations, the cumulative percentages along with the corresponding critical values are given in it (Appendix-X). For working out the crop-combination in an enumeration unit, one has to work out the percentage of each crop occupying upto one per cent of the total cropped area of the unit and arrange them in descending order. For ascertaining monoculture, one has to locate the percentage of the first ranking crop in the percentage column of monoculture and compare the critical value given against this percentage in the table with the percentage strength of the second crop. If the critical value is more than the percentage of the second crop, this crop is considered as insignificant crop and the unit is designated as monoculture area. But if the critical value of the table is less than the percentage of the second crop, this crop is considered as important crop and is included in the combination. Now to ascertain whether third crop is to be included in the combination or not, the percentages of first and second crops are added together and its position is located in the percentage column of 2 crop-combination and the critical value given against it is compared with the percentage of the third crop. If the critical value is more than the percentage of the third crop, this crop is considered as insignificant and the unit is identified as 2 crop-combinations.

Table 5.12: Crop-Combination Regions

	Districts	*Crop-Combinations*	
		2003-04	*2005-06*
1.	Uttarkashi	(5) WRMPV	(3) WRM
2.	Chamoli	(2) RW	(3) RWM
3.	Pithoragarh	(5) WRMPB	(4) WRMP
4.	Tehri Garhwal	(4) WMRP	(4) WMRP
5.	Rudraprayag	(3) WRM	(3) WRM
6.	Bageshwar	(3) WRM	(3) WRM
7.	Almora	(3) WMR	(3) WMR
8.	Garhwal	(3) WMR	(3) WMR
9.	Champawat	(4) WRMP	(4) WRMP
10.	Nainital	(5) WROVMa	(5) WROVMa
11.	Dehradun	(4) RWMS	(4) WRMaS
12.	Hardwar	(3) SWR	(3) SWR
13.	Udham Singh Nagar	(2) RW	(2) RW

Source : Compiled and Computed by the author.

*Figures in brackets indicate the number of crops in combination.

It can be illustrated by taking the example of US Nagar district where rice had occupied 42.46 per cent, wheat 35.87 per cent, sugarcane 10.63 per cent, oilseeds 3.60 per cent, pulses 1.40 and maize 1.09 per cent. Since, the first crop occupies 42.46 per cent and the critical value in monoculture column is zero in all cases of less than 51 percent; the second crop is significant. Now add the percentage of first and second crops (42.46 + 35.87 = 78.33) locate it in the column of 2 crop-combinations and compare the critical value given against it i.e. 20.0 with the percentage of third crop i.e. 10.63. Since the critical value is more than the third crop percentage, the crop is insignificant and the district is designated as 2 crop-combination region.

Similarly the position of subsequent crops is ascertained and crop-combinations are formed. Similarly it can be illustrated by taking the example of Chamoli district where wheat occupies 29.69 per cent, rice 22.49 per cent, madua 19.07 per cent, vegetables 4.84 per cent, pulses 3.60 per cent, barley 4.40 per cent and oil seeds 2.71 per cent.

Since the first crop occupies 29.69 per cent and the critical value is zero against all the percentages of less than 51 per cent, the second crop is significant. Add the percentages of first and second crops (29.69 + 22.49 = 52.18%) and locate it in the percentage column of 2-crop combination and find out the critical value against it which is 1.03. Since it is less than the percentage of third crop, this crop is significant. Now

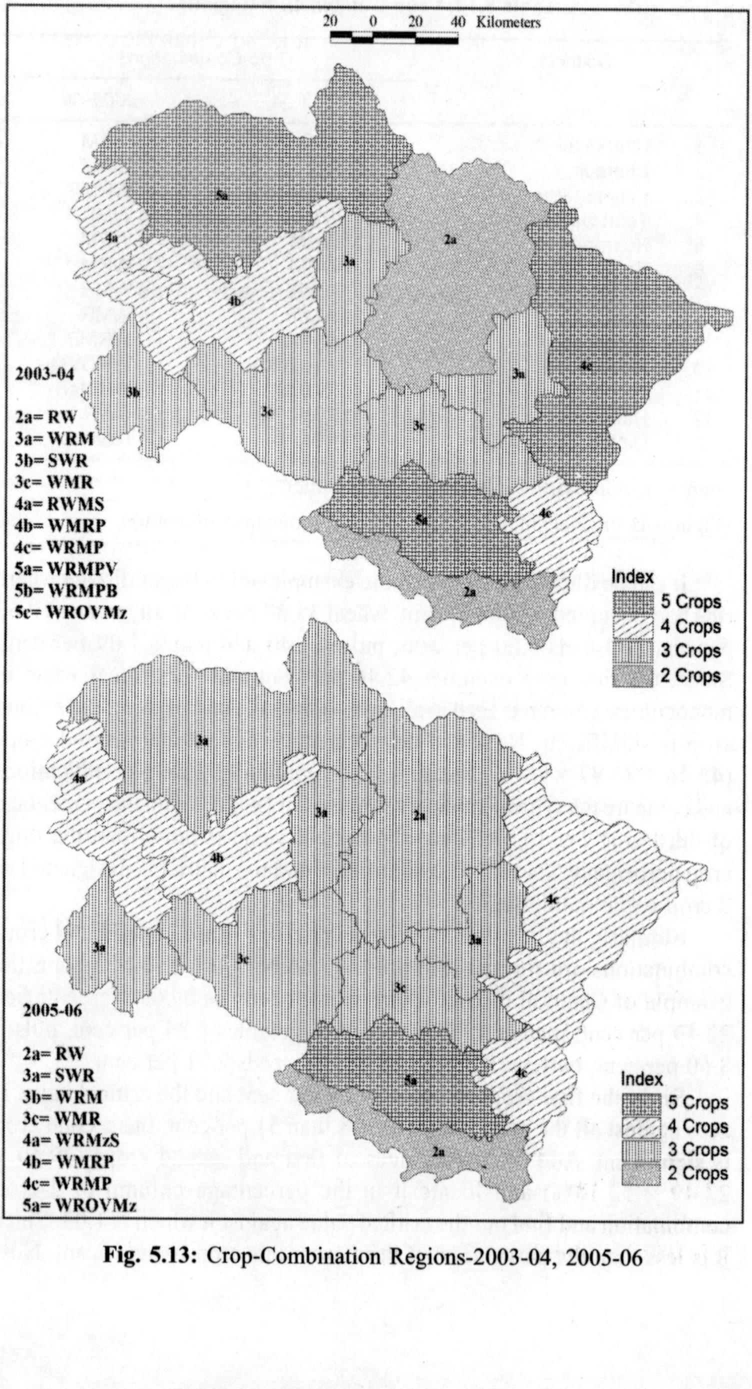

Fig. 5.13: Crop-Combination Regions-2003-04, 2005-06

add the percentage of first, second and third crops (29.69 + 22.49 + 19.07 = 71.25%) and located its position in the percentage column of 3-crop combination which is 8.42. Since, it is more than the percentage of the 4th crop i.e. 4.84, this crop is insignificant and the area is designated as 3 crop-combination (Appendix-X).

As a result of the statistical processing of the data based on Doi's method using Critical Table the crop-combinations in each districts have been worked out. The results are given in Table 5.12 and are shown on Fig. 5.13. It is obvious from the Table 5.12 that monoculture is non-existent in the state, and the regions from 2 crops to a combination of 5 crops have emerged. Altogether 8 crops have come in different combinations forming 8 regions of crop combinations. A brief description of each combination would be worthwhile.

2: Crop-Combination Region: It is only US Nagar district where 2 crops i.e. rice and wheat have come in combination in both the agricultural years. It is a district lying in the plain with favourable geographical conditions inhabited by enterprising Punjabi migrant farmers who prefer to get the maximum profit from their agricultural land using modern agricultural technology.

3: Crop-Combination Region: This is most common combination of crops found in 5 districts in 2003-04 and in 7 districts in 2005-06. It is found in a contiguous region spreading over 7 districts of Uttarkashi, Rudraprayag, Chamoli, Bageshwar, Almora, Garhwal and Hardwar having different combinations of four crops viz. wheat, rice, madua and sugarcane in 2005-06. There are six districts where wheat, rice and madua which are the most common crop in the state, have come in combination with their changing position, while in Hardwar wheat, rice and sugarcane have come in combination. Almost similar was the position in 2003-04.

4: Crop-Combination Region: This is the next common crop-combination found in 4 districts of Dehradun, Tehri Garhwal, Pithoragarh and Champawat in 2005-06, while it was found only in 3 districts of Tehri Garhwal, Champawat and Dehradun in 2003-04. As many as 6 crops viz. wheat, rice, madua, pulses, maize and sugarcane have come in combination with different associations in the districts. There are two sub-regions of this combination lying in the west including the districts of Dehradun and Tehri Garhwal, and in the east comprising the districts of Pithoragarh and Champawat in 2005-06. Of course the larger number of crops in combination indicates the diversification of agriculture. Almost similar was the status of crops in combination in 2003-04.

5: Crop-Combination Region: This crop-combination occurs in only one district of Nainital where wheat, rice, oilseeds, vegetables and maize have come in association in 2005-06. This is mainly because of its specific location and high level of urbanization requiring variety of crops. In 2003-05, 5: Crop-combination is found in 3 districts viz. Pithoragarh, Uttarkashi and Nainital.

5.6 Agricultural Diversification

During the last three decades or so, the chemical-irrigation based agriculture has no doubt increased the agricultural production but in has created and is continuously creating several environmental problems. Therefore, the Indian agriculture is heading towards diversification of agriculture into two ways : firstly, by diversifying the cropping pattern, and secondly, by diversifying the agricultural activity. Diversification of agriculture is preferred due to variety of reasons. Firstly, it is a kind of guarantee against the uncertainties and vagaries of monsoon; secondly, it prevents the deterioration and degradation of land because some crops are nitrogen fixing, while some generate bacteria in the soil; thirdly, it provides jobs to the farmers all the year round; fourthly it requires lesser capital input; and finally, it provides quality food having various elements of diet in adequate amount (Chand, 1999; Mirthunjaya, 2002 and Jana, 2006).

There are various methods of measuring crop-diversification (Martin, 1962; Bhatia, 1960 and 1965 and Singh, 1976). In all these methods the percentage strength of all the crops with reference to gross cropped area is calculated. Gibs-Martin has evolved the following formula for working out the index of crop diversification :

$$\text{ICD} = \frac{\sum x^2}{\left(\sum n\right)^2}$$

where

ICD = Index of Crop Diversification
x = Percentage of different crops to gross cropped area
n = Number of crops

In this formula the value of CDI is 0.9 in case of maximum diversification of crops, while it is zero in case of monoculture.

Based on the percentage strength of the crops Bhatia has developed the following formula for assessing the crop diversification.

$$\text{ICD} = \frac{\sum_{i=1}^{n} x_i}{n}$$

where

ICD = Index of Crop Diversification
x_i = Percentage of crops to gross cropped area
n = Number of crops occupying 10 per cent or more of the gross cropped land

Bhatia has taken only those crops which occupy upto 10 per cent of the gross cropped area. Singh, 1972 using the same percentage strength of the crop has modified this formula by taking all the crops occupying 5 per cent or more of the gross cropped area. In both the cases inverse is the relationship between the crop concentration index and crop diversification. Lower the value of index, higher is the crop diversification and *vice-versa*.

The author has further modified Bhatia's formula of crop diversification and has considered all the crops occupying one per cent or more of the gross cropped area. The author, using his own modified formula, has worked out the crop diversification index for all the districts. Based on the results thus obtained, the districts have been arranged in ascending order of crop diversification index as lower the value of index, higher is the crop diversification and vice-versa and have finally been grouped into four categories (Table 5.13).

Table 5.13: Level of Agricultural Diversification : 2005-06

Level of agr. div.	*ICD range*	*No. of dist.*	*%age to total*
Very High	< 7.00	2	15.40
High	7.00- 9.99	4	30.70
Moderate	10.00- 13.00	3	23-20
Low	> 13-00	4	30-70
		13	100-00

Source : Compiled and Computed by the author.

The level of agricultural diversification has been represented in Fig. 5.14 which shows that very high level of agricultural diversification is found in the districts of Nainital and Dehradun which have very high level of urbanization requiring variety of agricultural produce. Therefore, the farmers grow vegetables, fodder, oilseeds, besides rice, wheat and madua which give very high level of agricultural diversification.

High level of agricultural diversification is found in 4 districts including Uttarkashi, Tehri Garhwal, Almora and Champawat. The former

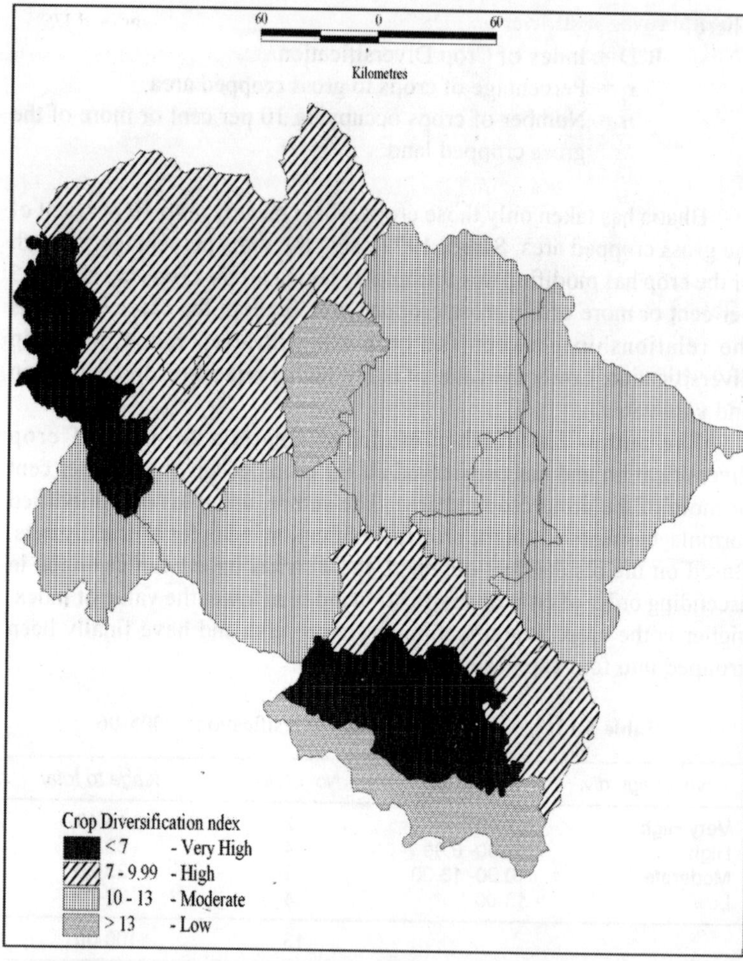

Fig. 5.14: Crop Diversification : 2005-06

two districts forming a contiguous belt lie in the northwestern part of the state, while the latter two districts lie in the eastern part, and form a single region. The region of moderate level of agricultural diversification covering 4 districts of Chamoli, Pauri Garhwal, Bageshwar and Pithoragarh constitutes a single region substantially located in the northeastern part of the state with the exception of Pauri Garhwal district. The region of low level of agricultural diversification with crop diversification index of more than 13, has been found in 3 districts of

Rudraprayag, Hardwar and US Nagar. The districts of Hardwar and US Nagar are agriculturally most advanced districts with high level of agricultural technology. Here, agricultural is highly commercialized with the specialization in rice, wheat and sugarcane cultivation. In Rudraprayag the low level of agricultural diversification has emerged, perhaps, by chance because here several crops are grown due to varied physical conditions.

References

Ahmad, M.S., 2004 : *Five Decades of Planning and Developing : A Case Study of Uttaranchal and Jharkhand*, Gyan Pal House, New Delhi.

Alam, Z, 2003: Sustainable Agricultural Development in Nainital District (Unpublished M.Phil. Dissertation, Dept. of Geog., Delhi University).

Athewala, A.G., 1966 : Some New Methods of Crop-Combination, *Geographical Review of India*, Vol. XXXVIII (1), Kolkata.

Ayyar, N.P., 1969 : Crop Regions of Madhya Pradesh, *Geographical Review of India*, Vol. XXXI (1), Kolkata.

Bandooni, S.K., 2004 : *Land Resource Management and Development in Hill Areas*, Research India Press, New Delhi.

Bhatia, S.S., 1960 : An Index of Crop Diversification, *Professional Geographer*, Vol. XXII.

Bhatia, S.S., 1965. *Patterns of Crop Concentration and Diversification and Development of Mountain Regions with Special Reference to Himachal Pradesh*, M.D. Pub. Pvt. Ltd., New Delhi.

Chand, R., 1999 : *Agricultural Diversification in India : Potential and Prospects*, Mittal Pub., N. Delhi.

Doi, K., 1959 : The Industrial Structure of Japanese Prefectures, *Proceedings of the IGU Regional Conference in Japan, 1957*, Japan.

Jana, B.L., 2006 : *Diversification in Agriculture*, Sanjanya Book, Delhi.

Jodha, N.S. *et al.*, (Eds.), 1992 : *Sustainable Mountain Agriculture : Perspectives and Issues*, Oxford and IBH, New Delhi.

Johnson, B.L.C., 1958 : Crop Association Regions in East Pakistan, *Geography*, Vol. XLII.

Joshi, S.C., 2004 : *Environment and Development : Geo-ecological Overview*, Gyanodaya Pub., Nainital.

Kostrowicki, J., 1972 : 'Some Methods and Techniques to Determine Crop and Other Landuse Combinations as used in Polish Landuse Studies' in Shafi, M., Anas, M. and Siddiqui, M.F. (Eds.), *Proceeding of Symposium on Landuse in Developing Countries*, Deptt,. of Geography, Aligarh.

Mahmood, A., 1977 : *Statistical Methods in Geographical Studies*, Rajesh Publications, New Delhi.

Maikhuri, R.M. *et al.*, 2004 : 'Agriculture is Uttarakhand : Issues and Management Prospects for Economic Development' in Sati, M.C. and Sati S.P. (Eds.), *Uttarakhand State : Dimensions of Development*, Indus Pub., New Delhi.

Majeed, A., 1992 : Crop-Combinational Analysis : A Review in Methodology' in Mohammad, N. (Ed.), *New Dimensions in Agricultural Geography*, Vol.IV, Concept Pub. Co., New Delhi.

Masood, S., 2009 : *Peri-Urban Agriculture : Agriculture by Twenty 20 and Beyond*, APH Pub. Corp., New Delhi.

Mehta, G.S., 1999 : *Development of Uttarakhand : Issues and Perspectives*, APH Pub. Corp., New Delhi.

Mohammad, N., 1975 : Crop Association Regions in Ghaghara-Rapti Doab, *The North Eastern Geographer*, Vol.VII (182).

Mohammad, N., 1992 : *The Ecology of Agricultural System*, Vol. II, Concept Pub. Comp., New Delhi.

Nag, P. (Ed.), 2010: *Atlas of Agricultural Resources of India*, National Atlas and Thematic Mapping Organization, Kolkata.

Nelson, H.J., 1955 : A Service Classification of American Cities, *Economic Geography*, Vol. XXXI (3).

Pownall, L.L., 1953. The Functions of New Zealand Towns, *Annals of the Association of American Geographers*, Vol.43.

Rafiuallah, S.M., 1965 : A New Approach to Functional Classification of Towns, *The Geographer*, Vol. XXII, Aligarh.

Siddiqui, M.F., 1975 : Crop-Combination and Specialization in India, *The Geographer*, Vol. XXI, Aligarh.

Singh, J. and Dhillon, S.S., 1984 : *Agricultural Geography*, Tata McGraw Hill Pub. Co., New Delhi.

Singh, J.S., 1985 : *Environmental Regeneration in Himalaya*, Gyanendra Parkashan, Nainital.

Weaver, J.C., 1954 : Crop-Combination Regions in the Middle West, *Geographical Review*, XLIV (2).

Chapter 6

Agricultural Diversification and Food Security
A Meso Level Analysis

6.1 Introduction*

Food security has become the buzz word of the day. It has been a matter of great concern to academicians, politicians, planners and decision makers all over the world during the last three decades due to the unprecedented growth of population. The present world population of 6 billions is likely to reach 8 billion at the end of first quarter of this century. Each year there will be approximately 90 million additional mouths to feed and bodies to clothe, and about 90 per cent of this increase would be in the low income countries. According to an estimate about 800 million people already suffer from hunger, and approximately 1.3 billion people subsist on an income of less than one US dollar per day. The World Bank has estimated that 1,115 million people were below the poverty line and out of these 630 million suffered from extreme poverty. It further stated that 80 per cent of the world's poor live in rural areas where bulk of the people earns their living from farming (Chopra, *et al.*, 1998). But the status of farming on the whole is not encouraging either. Recent FAO assessment indicates that world agricultural production including both crops and livestock, is going down in comparison to 1980s and the growth rate of foodgrains output in 1990s was almost half, i.e. 1.8 per cent against 3.54 per cent (Murlidharan, 2001). Among the developing countries as a

* The introductory part of this chapter is taken from my own article 'Regional Pattern of Food Security in India", in Singh, T. (Ed.) 2004 : *Resource Conservation and Food Security : An Indian Experience*, Concept Pub. Co., New Delhi.

whole, agricultural production of 1997 has experienced the lowest increase since 1979, and has barely kept pace with population growth, while in developed countries overall agricultural output slowed down in 1997, and it is not expected to be able to meet the anticipated consumption requirements, and the global stocks are to be drawn down (Luppin and Menzol, 1999). Therefore, food security has been a persistent issue raised in several national and international seminars and conferences organized during the last two decades. It has also found a prominent place in the research themes of scholars specially the geographers, economists and sociologists (Minhas, 1976; Sen, 1981; Borlaug, 2004; Radhakrishna, 1991; Tendulkar, *et al.,* 1993; George, 1996; Suryanarayana, 1996; Masood, 2009, and Scanes and Miranowski, 2004).

The World Food Conference of 1974 held at Rome, which was organized in the wake of 1972-74 world food crisis, made a call for ending hunger by 1984 which led to the enunciation by FAO of the concept of food security defined as physical and economic access to food to all people at all times. In 1976 Minhas in his Presidential address to 36th Annual Conference of the Indian Society of Agricultural Economics has cautioned the politicians, people and the academicians about the issue of food security in India. The UN Conference on Desertification held at Nairobi in 1977 alerted political leaders and the public to the serious threats to both food and livelihood security arising from the growing deterioration in the biological potential of the soil as a result of anthropogenic pressure and unsustainable exploitation. The 1992 Rio de Jeneiro Conference on Environment and Development emphasized the need to ensure food security at all levels within the framework of sustainable development. In the International Conference on Population and Development held at Cairo in 1994 the linkages between population growth and food production and the need to evolve global measures to satisfy the ever growing food needs was the focal theme of discussion. In 1995 the World Summit on Social Development held at Copenhagen made a strong commitment to campaign against hunger through its emphasis on poverty eradication. Similarly, the World Food Summit held at Rome in 1996 gave a new impetus to fight for food security by focusing attention on food issues. In 2001 a National Consultation with the focal theme Towards Hunger Free India was organized in Delhi with the objective of evolving an action plan and policies that would ensure sustainable food security at the national level and the elimination of poverty induced chronic hunger at the household level by the end of the Tenth Five Year Plan (Bader, *et al.,* 2010). Two documents, viz. "Food Security Atlas of

Rural India" and "Enabling Development—Food Assistance in South Asia" have been released. Food Security Atlas of Rural India (Swaminathan, 2001) highlights the food security as the story of poverty, discrimination and neglect of livelihoods, unhygienic living, lack of literacy, basic amenities and health care which is the indicator of failed global and national governance. Three dimensions of food insecurity, viz. availability, access and absorption have been highlighted. Food absorption–the ability of an individual to assimilate and absorb the food consumed depends on the status of health.

6.2 Concept of Food Security

The concept of food security is very complex, multidimensional and complicated and is debated since last three decades. It has been visualised at international, national, regional, household and even at individual levels. The World Food Conference of 1974 was mainly concerned with global food security and it has realised that world food crisis was a common responsibility of all nations and that international approaches were required to achieve world food security (Sarris and Taylor, 1976). Fluctuations in food supplies over time and unstable food prices are the basic problems of food security which require international and national solutions such as inter-regional grain reserves of buffer stocks, grain insurance and others (Johnson, 1976 and Valdes , 1981). Food security is generally understood to imply arrangements whereby the people would be assured of a minimum level of foodgrains supply in year of normal as well as poor harvest, which implies the physical supply of a minimum level foodgrains during all periods including those of harvest failures (Reutlinger, 1977). Subsequently, it was realized that physical availability alone would not ensure economic access to food for all population but satisfactory production levels and stability of supplies should be matched by reduction in poverty and an increase in the effective demand to ensure economic and physical access for the poor (FAO, 1987). As such the concept of food security at the individual and household levels was taken up (Frankenberger and McCarton, 1998). Hence, according to FAO (1984) the basic concept of food security implies that "all people at all times have both physical and economic access to the "basic food they need" but the World Bank (1986) has modified it indicating that "food security is access by all people at all times to enough food for an active and healthy life." Its essential elements are availability of food and the ability to acquire it. The World Bank has made a distinction between chronic and transitory food insecurities. The former refers to persistent inadequate diet caused by the inability to acquire food either by purchase or by produce of their own, while the latter refers to a temporary decline in the household's access to enough food due to instability in food prices, food production and household income. However, the most comprehensive and perhaps, largely acceptable definition came out of the World Food Summit at Rome in 1996. Food security exists when all people, at all times, have physical and economic access to sufficient, safe and nutrition food to meet their dietary needs and food preferences for an active and healthy life. This definition emphasizes three critical dimensions of food security : (i) physical supply of desired food in sufficient quantity; (ii) an economic access indicating close link of purchasing power to food security; and (iii) stability in supply which include access to global food market (Bhattacharya, 2001). A more comprehensive explanation of food security includes livelihood security for each household ensuring both physical and economic access to balanced diet, safe drinking water, environmental sanitation, primary education and basic health care (George, 1999). Finally, I would like to quote Chung, *et al.,* (1997) who has summarized the diverse determinants of food security status in a general conceptual framework focusing on the links between resources commanded by the household, levels of farm and non-farm production, household income, household and individual consumption and individual nutrition. The above review of the studies and the explanation given clearly reveals that the concept of food security is complex and multidimensional.

6.3 Indicators of Food Security

As the concept of food security is complex and multidimensional, and its various variables are interlinked with one another, it is difficult to evolve a single indicator for it. Chung, *et al.,* (1997) have taken a number of socio-economic indicators such as owning poor quality land or no land, holding distress sales of livestock productive and valued assets, taking loans, substitution of inferior quality staple food and vegetables, migration for search of work, working of pregnant and lactating mothers, etc. for measuring household food insecurity. Barun, *et al.,* (1992) have suggested demand and supply of food at country level, direct comprehensive surveys of dietary intake at household level and anthropometric information at individual level to measure the food security. In some studies poverty has been taken as the main determinant of chronic food insecurity, as most of the qualitative and quantitative indicators of food security are linked to poverty issues. The poor do not have adequate means or entitlements to secure their access to food, even if it is available in local and regional markets (Sen, 1981). Hence, the household income contributes to the access of food but it may not improve the nutritional level of all the individuals of the households. Hendry, agreeing the poverty as most effective indicator of food security has added that the extent of access to gainful employment, arable land, suitable technologies and other productive resources are other important factors to be reckoned within this context. The above review of all these studies and suggested indicators clearly indicates the complexity of selection of precise indicator and methodology to measure food security at larger scale.

6.4 Data Base and Methodology

In this research work food security has been discussed and analysed at three levels viz. district level (Meso Level), Village and household levels (Micro level). The meso level analysis is entirely based on secondary data, while the micro level analysis is based on primary data. The former is confined to demand and supply data.

In the absence of required data at district level for the state as a whole to measure it more rationally and scientifically the study is confined to demand and supply indicator. In order to ascertain the demand of food for the inhabitants, census data have been taken, while for ascertaining its supply the data published by Directorate of Economic and Statistics, Ministry of Agriculture have been tapped. As far as I know, very few studies have been done to ascertain the food security/insecurity at district level in India. Swaminathan (2001) has made a maiden attempt to study this issue but at the state level. Realising the problem of getting precise export and import data of all agricultural products at district level, a direct method developed by the author for measuring food security or insecurity is used which is based on the computation of net food available in terms of calories and monetary value and the consumption by the weighted total population with allowances for wastage, seed and feed (Mohammad, 2003). This approach has been adopted because the direct estimate of production and consumption is the only reliable way to ascertain food security in a unit area, and to compare one area with another.

The triennium average for the years 2003-04, 2004-05 and 2005-06 have been taken of the projected total population with the simple reason that the latest agricultural data at district level for the whole state is available upto 2005-06. As this average population data consists of the persons belonging to different demographic and socio-economic condition, but the requirement of food either in weight or calories varies according to age, sex, occupation, body size, income, culture and climate. Hence, there is a need of a method to standardise the population data. Aykroyd's (1956) co-efficient has not been taken here firstly because it does not give a co-efficient for children under 4 years of age, although substantial percentage of this age group of population depends on diets, and secondly, he used the same coefficient for adult female and children of 12 to 13 years of age (Chakraborty, 1970). Therefore, Singh's (1948) scale of consumption unit has been taken in this study. According to Singh's scale each unit of the population is equal to 0.773 consumption unit or 1000 persons are equal to 773 consumption units. Therefore, the total population of each district has been multiplied by the co-efficient of consumption unities 0.773 in order to get the consumption unit of the total population. Chakraborty has tested it in several districts of India selected on random basis and has found it reasonably sound enough to be used as consumption unit.

Food availability has been worked out at district level by taking triennium average data of agricultural produce for the years 2003-04, 2004-05 and 2005-06. There are many sources of leakage, e.g. losses in transport and storage, damage by insects and pests, kitchen waste, use of seeds etc., between the production and consumption of agricultural produce but the precise data for all these aspects are not available. Besides these, certain percentage of the net sown area is given to fodder crops in India. Therefore, Chakravarty (1970), after examining various estimates of losses on these accounts given by different scholars and institutions has suggested the standard deduction of 10 per cent exclusive of seed requirement. Similarly after evaluating various estimates of seed requirements ranging from 3.5 to 14.0 per cent, he has recommended 6.8 per cent of the total production for seed requirements. Hence, the total deduction for wastage, cattle feed and seed comes out to be 16.8 per cent of the total gross production. The co-efficient of production, therefore, is 0.832 (100-16.8 = 83.2). The availability of food, in the present study has been worked out at two levels, i.e. availability of food in terms of calorific value and monetary value which ascertains the access to food. For the former all the agricultural produce and animal products such as rice, wheat, bajra, ragi, jowar, maize, barley, gram, other pulses, vegetables and fruits, sugarcane, etc. which could be converted into calories have been considered. As the nature of all the produce varies, they cannot be aggregated in terms of weight. Therefore, they have been converted into calorific value with help of conversion table and have been added together to get the food production in calories for each of the district (Gopalan, *et al.*, 2002). The resulting figures have been multiplied by the co-efficient of production, i.e. 0.832 in order to get the net food available in terms of calories.

But in India agriculture is highly diversified and the crops ranging from opium to rice are grown in different parts of the country. The nature and character of the crops grown is as such that they cannot be added or aggregated together. Hence, various methods such as weight, foodgrain equivalence, standard nutrition unit, calories and monetary value have been suggested (Singh, 1972; Kendal, 1939; and Raza, 1992). In the present study monetary value, although it also has many problems of selection of prices due to its seasonal and annual fluctuation, has been taken up as the other methods have several limitations (Mohammad, 1992). Thus the total agricultural produce has been converted into monetary value by multiplying the average production of crops with their state level average prices. The result, thus obtained, has been multiplied by the coefficient of production, i.e. 0.832 in order to get net availability of agricultural produce in monetary value which ensures access to food within household, local, regional or national markets.

6.5 Spatial Pattern of Food Security

The spatial pattern of food security in Uttarakhand has been ascertained for the total population based on total agricultural production in terms of calories and monetary value. The entire theme has been discussed under the following system :

- 6.5.1 District-wise consumption units.
- 6.5.2 Spatial pattern of consumption units
- 6.5.3 Spatial pattern of carrying capacity of land
  - 6.5.3.1 Carrying capacity of land in calorific value
  - 6.5.3.2 Carrying capacity of land in monetary value
- 6.5.4 Spatial pattern of food security
  - 6.5.4.1 Spatial Pattern of food security in calorific value
  - 6.5.4.2 Spatial pattern of food security in monetary value

6.5.1 District-wise Consumption Unit

Like the nation as a whole, the demographic structure of the state is varied and complex consisting of people of different age, sex, occupation, etc.

However, the triennium average population of the projected total population for the years 2003-04, 2004-05 and 2005-06 of each district has been converted into consumption units by multiplying it with co-efficient of consumption unit i.e., 0.773 for the simple reason that the latest agriculture data at district level for the whole state are available upto 2005-06. The results thus obtained have been given in Table 6.1 which shows a lot of district-wise variations in it. It has been represented in Fig. 6.1. It ranges from as low as 2.6 per cent of the total in district Champawat to as high as 18 per cent in the district of Hardwar. It may be observed that the districts lying in tarai plains or in major river valley basins have higher percentage of total consumption units. The Garwal division accounts nearly 58 per cent of the total consumption units, while remaining 42 per cent lies in the Kumaon division.

Table 6.1: Uttarakhand: District-wise Consumption Units-2004-05

District	*Average project pop. 2003-04 to 2005-06*	*Consumption units*	*%age of cons. units*
Uttarkashi	3,12,850	2,41,833	3.35
Chamoli	3,80,937	2,94,464	4.08
Rudraprayag	2,33,153	1,80,227	2.50
Tehri Garhwal	6,10,862	4,72,196	6.55
Dehradun	14,92,349	11,53,586	16.01
Pauri Garhwal	6,91,803	5,34,764	7.41
Hardwar	16,89,113	13,05,685	18.10
Garhwal Div.	**54,11,067**	**41,82,755**	**58.00**
Bageshwar	25,54,401	1,97,425	2.74
Almora	6,26,336	4,84,157	6.71
Champawat	2,43,730	1,88,403	2.61
Nainital	8,62,028	6,66,348	9.24
US Nagar	14,55,992	11,25,482	15.51
Pithoragarh	14,75,248	3,67,367	5.09
Kumaon Div.	**39,18,735**	**30,29,182**	**42.00**
Total Uttrakhand	**93,29,802**	**72,11,937**	**100.00**

Source: Compiled and Computed by the author.

6.5.2 *Spatial Pattern of Consumption Units*

The preceeding discuss shows a lot of variations in the consumption units of the districts. Therefore, based on these variations, the districts have been grouped into three categories i.e. high concentration, moderate concentration and low concentration (Table 6.2).

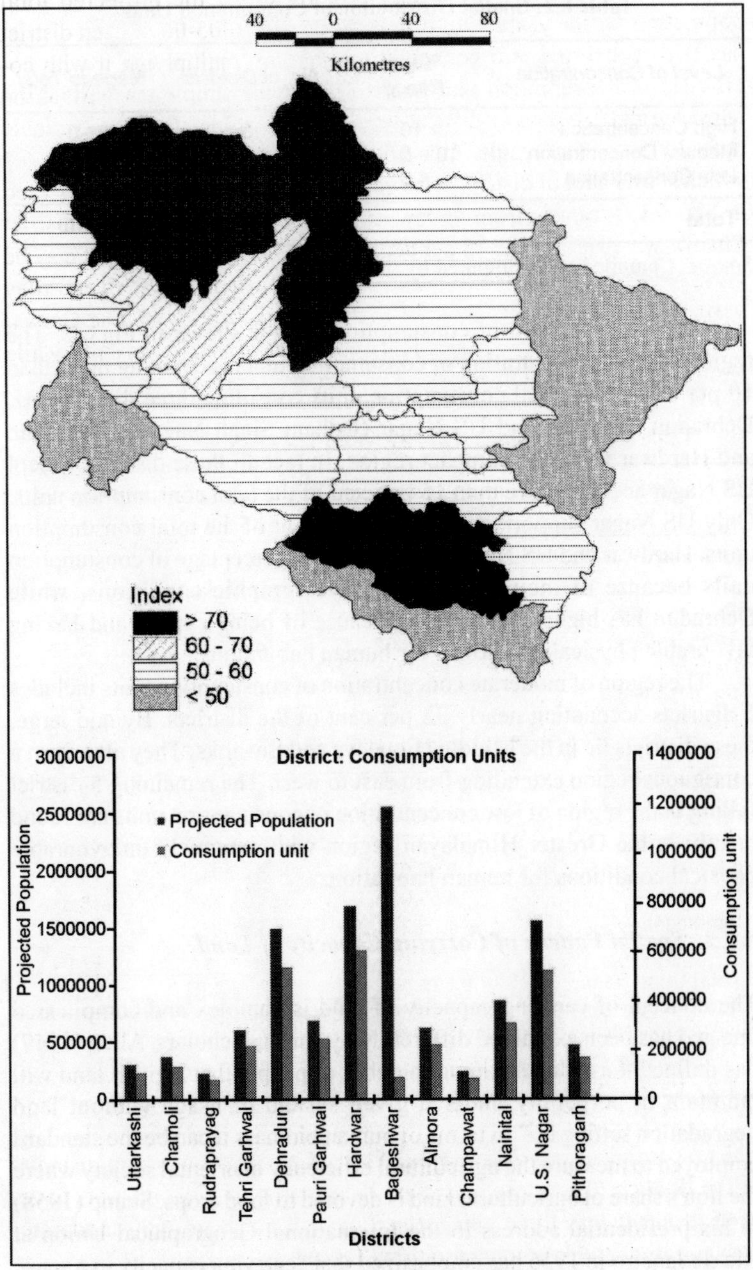

Fig. 6.1: Distribution of Consumption Units

Table 6.2: Spatial Distribution of Consumption Units

Level of Concentration	*%age Range*	*No. of Districts*	*%age to Total*
High Concentration	> 10	3	23-0
Medium Concentration	10 – 5	5	38.5
Low Concentration	< 5	5	38.5
Total		**13**	**100.00**

Source: Compiled and Computed by the author.

The spatial pattern of consumption units is shown in Fig.6.2. The region of high concentration of consumption units accounting more than 10 per cent of the total consumption units constitute three districts viz. Dehradun, Hardwar and US Nagar (Udham Singh Nagar),. Dehradun and Hardwar form one compact region. In fact all these districts except US Nagar account more than 15 per cent of the total consumption units. Only US Nagar supports nearly 14.6 per cent of the total consumption units. Hardwar and US Nagar support higher percentage of consumption units because of their favourable physiographic conditions, while Dehradun has higher percentage because of being capital and having favourable physical conditions for human habitations.

The region of moderate concentration of consumption units includes 5 districts accounting nearly 38 per cent of the districts. By and large, these districts lie in the Middle Himalaya and Siwaliks. They also form a contiguous region extending from east to west. The remaining 5 district falling in the region of low concentration of consumption units are found mostly in the Greater Himalayan region with extremely unfavourable physical conditions for human habitation.

6.5.3 Spatial Pattern of Carrying Capacity of Land

The concept of carrying capacity of land is complex and complicated one and has been explained differently by various scholars. Allan (1949) has defined it as "the maximum number of people that a given land will maintain in perpetuity under a given system of usage without land degradation setting in." In terms of human biomass it can be the standard employed to measure the agricultural efficiency in oriental society where the lion's share of agricultural land is devoted to food crops. Stamp (1958) in his presidential address to the International Geographical Union at Rio de Janeiro in 1956 has emphasized that "carrying capacity in a sense is a measure of farm efficiency". He has employed two parameters viz. Standard Nutrition Unit and Calorific value to measure the carrying capacity of land. He has also developed the concept of Potential Production Unit (PPU) for measuring carrying capacity of land. However, the common denominator of all these definitions and explanations is that carrying capacity of land is defined as the number of persons that a unit area of land can support at a given technological level and system of landuse.

There are several methods for measuring the carrying capacity of land (Stamp, 1958; Khusro, 1965; Shafi, 1972; Gopalan, 2000; Chakravarty, 1970; Bernard and Derrik, 2007 and Singh, 1970). Stamps has developed the concept of Potential Production Unit (PPU) for measuring the land potentiality in which the average yield per unit of land of the crop of the medium quality land has been taken as one PPU. Based on it and the yield of good quality, medium quality and poor quality lands, the PPU has been worked out. For example, if the average yield of good quality land is 20 quintal, medium quality is 10 quintal and poor quality land is 5 quintal per ha, one hectare of good quality land is equal to 2 PPU, while one hectare of poor quality land is equal to 0.5 PPU. Stamp has also measured the carrying capacity of land in terms of calorific value and has developed the concept of Standard Nutrition Unit (SNU). Shafi has widely used it with certain modifications. In this method firstly the entire food crops are converted into calories by using the standard conversion scale (Gopalan, 2000), secondly, the standard requirement of calories per head per day is ascertained which is multiplied by 365 in order to get the requirement of calories per head per annum. Finally 10 per cent of it is added to annual requirement of calories to compensate the wastage in the processing etc. The result thus obtained is termed as SNU. At the end the total calories produced in one unit of land is divided with the value of SNU in order to get number of SNU's produced in one unit of land. The result thus obtained will indicate that one unit of land can support so many persons which is the carrying capacity of land.

In the present study the concept of SNU has been applied with slight modification. The carrying capacity has been worked out in two units viz. (I) Calories and (II) Monetary Value.

6.5.3.1 Carrying Capacity of Land in Calorific Value

In this method firstly all agricultural products have been converted into calorific value which gives the gross availability of food. It has been multiplied by the co-efficient of production (0.832) in order to get the net availability of food which has been divided by the net area sown in order to get the net available food per hectare of land. The result thus obtained is divided by the standard requirement of calories per head per annum which gives the carrying capacity of land. The standard requirement has been ascertained by multiplying the standard requirement of calories (2,430) by 365 (2,430 × 365 = 8,86,950). About 12.5 per cent of it as processing wastage and seed has been added to it which comes out 997818. It has been round to 10,00,000 to be considered as one Standard Nutrition Unit.

Based on this methodology all agricultural products convertible into calories have been converted separately into calories for each district using standard conversion table and have been added together which gives the gross food available in calories. It has been multiplied by the co-efficient of production (0.832) in order to get the net availability of food which has been divided by the net sown area of the district that gives the net availability food in calorific value per unit of land i.e. hectare. It has further been divided by the standard requirement of calories per head per annum (10,00,000) which gives the carrying capacity of land i.e. one hectare of land can support how many persons in the district. The results thus obtained have been presented in Table 6.3.

Table 6.3: District-wise Carrying Capacity of Land in Calorific Value

District	*Per ha net food available (Calories)*	*No. of S.N.U.*
Uttarkashi	64,85,961	6.49
Chamoli	69,75,852	6.98
Rudraprayag	79,29,988	7.93
Tehri Garhwal	81,59,547	8.16
Dehradun	2,61,28,723	26.13
Pauri Garhwal	57,11,280	5.71
Hardwar	1,22,47,325	12.25
Bageshwar	73,89,617	7.39
Almora	50,15,867	5.02
Champawat	70,02,940	7.00
Nainital	1,48,39,570	14.84
US Nagar	8,92,572	8.93
Pithoragarh	57,97,051	5.80

Source : Compiled and Computed by the author

Table 6.3 shows a lot of variations in the net availability of calories ranging from as low as 50,15,867 per hectare in Almora to as high as 2,61,28,723 calories in Dehradun. Consequently, there are variations in

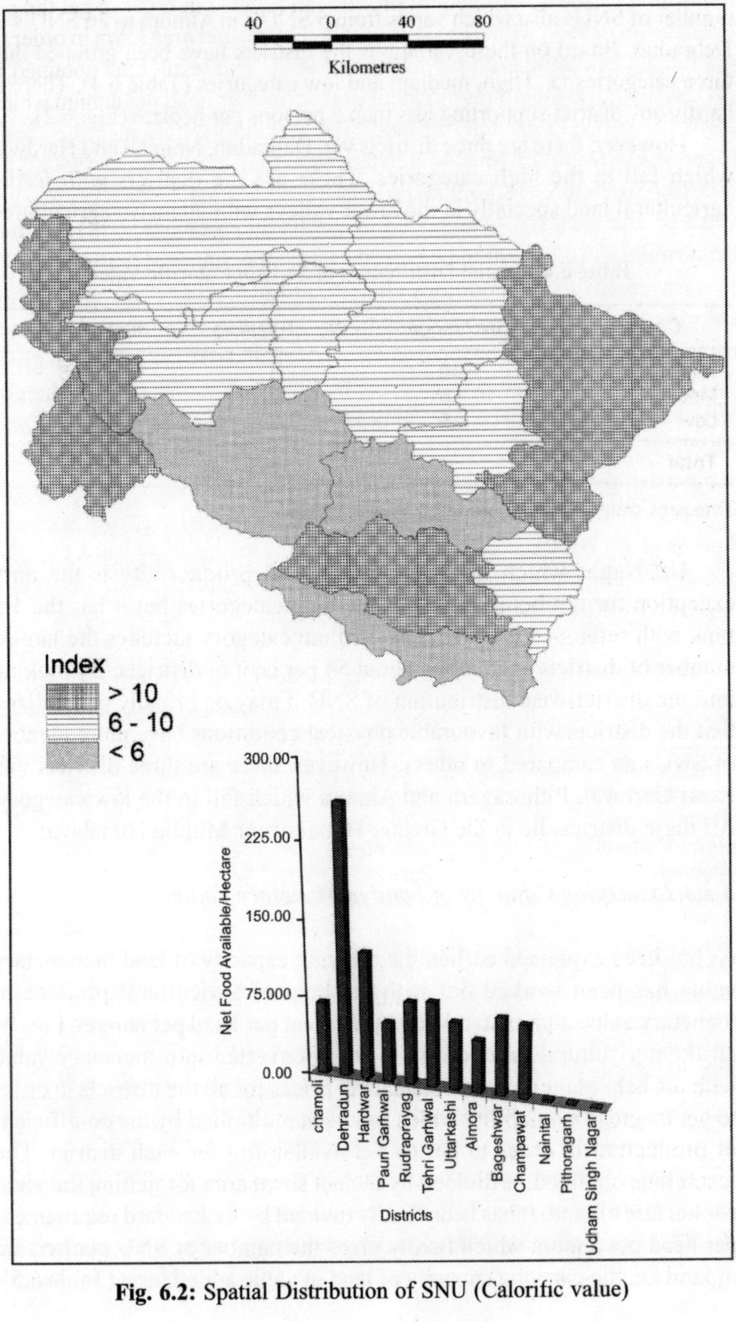

Fig. 6.2: Spatial Distribution of SNU (Calorific value)

number of SNUs also which varies from 5 SNU's in Almora to 26 SNU's in Dehradun. Based on these variations the districts have been grouped into three categories i.e. High, medium and low categories (Table 6.4). There is hardly any district supporting less than 5 persons per hectare (Fig. 6.2).

However, there are three districts viz. Dehradun, Nainital and Hardwar which fall in the high categories. These are the districts with fertile agricultural land specially in the lower valleys with intensive agriculture.

Table 6.4: Spatial Distribution of SNU in Calorific Value

Category	*SNU range*	*No. of districts*	*%age of total*
High	> 10	3	23.08
Medium	10 – 6	7	54.84
Low	< 5	3	23.08
Total		**13**	**100.0**

Source: Compiled and Computed by the author

US Nagar which has high agricultural productivity is the only exception for not being falling in the high categories but it has the 4th rank with reference to SNU. The medium category includes the largest number of districts accounting about 54 per cent of districts. By looking into the district-wise distribution of SNU it may be broadly generalized that the districts with favourable physical conditions have more number of SNUs as compared to others. However, there are three districts viz. Pauri Garhwal, Pithoragarh and Almora which fall in the low category. All these districts lie in the Greater Himalaya or Middle Himalaya.

6.5.3.2 Carrying Capacity of Land in Monetary Value

As has been explained earlier, the carrying capacity of land in monetary value has been worked out with the help of agricultural produce in monetary value and its standard requirement per head per annum. Firstly, all the agricultural products have been converted into monetary value with the help of their current state level prices for all the districts in order to get its gross availability which has been multiplied by the co-efficient of production in order to get its net availability for each district. The result thus obtained, is divided by the net sown area for getting the yield per hectare of land. It has been finally divided by its standard requirement per head per annum which finally gives the number of SNU per hectare of land i.e. the carrying capacity of land in value added term (Table 6.5).

Table 6.5: District-wise Carrying Capacity of Land in Monetary Value

District	*Per ha Net Food Available (Calories)*	*No. of S.N.U.*
Uttarkashi	70,481	6.41
Chamoli	89,235	8.11
Rudraprayag	1,09,128	9.92
Tehri Garhwal	1,14,439	10.40
Dehradun	4,19,574	38.14
Pauri Garhwal	95,611	8.69
Hardwar	1,76,111	16.01
Bageshwar	92,050	8.37
Almora	66,549	6.05
Champawat	82,051	7.46
Nainital	2,54,156	23.11
US Nagar	1,54,447	14.04
Pithoragarh	60,785	5.53

Source: Compiled and Computed by the author

Table 6.5 shows a lot of variations in the net availability of monetary value and consequently in the standard nutrition unit per hectare of net sown area. The highest availability is found in Dehradun (Rs. 4,19,574 ha) followed by Nainital (Rs. 2,54,156, whereas the lowest is recorded in Pithoragarh (Rs. 6,875) followed by Almora (Rs. 66,549 per head per annum). Similar is the case with availability of SNU i.e. carrying capacity of land (Fig. 6.4). However, based on these variations in carrying capacity of land in different districts, they have grouped in 3 categories i.e. region of high carrying capacity, medium carrying capacity and low carrying capacity (Table 6.6).

The spatial distribution of carrying capacity of land in value added term is represented in Fig. 6.3 which shows that the region of high carrying capacity of SNU including 5 districts, lies in the southern part of the state. It consists of two sub-regions being separated by Pauri Garhwal district. The larger sub-region consisting of Dehradun, Hardwar and Tehri Garhwal lies in the northwestern part of the state. Dehradun has the highest carrying capacity in the state. The reasons for high SNU in Dehradun are its being the capital with high population creating more demand for horticultural products and the vicinity to the plain leading to more accessibility. In all these districts the commercial agriculture specially horticulture which is comparatively more remuneration is the main cause of high number of SNU. Obviously the favourable physiographic conditions specially relief, climate and soil are the other causes for it (Fig. 6.3).

The map further shows that there are 7 districts accounting nearly 54 per cent of the total districts fall in the region of medium carrying capacity.

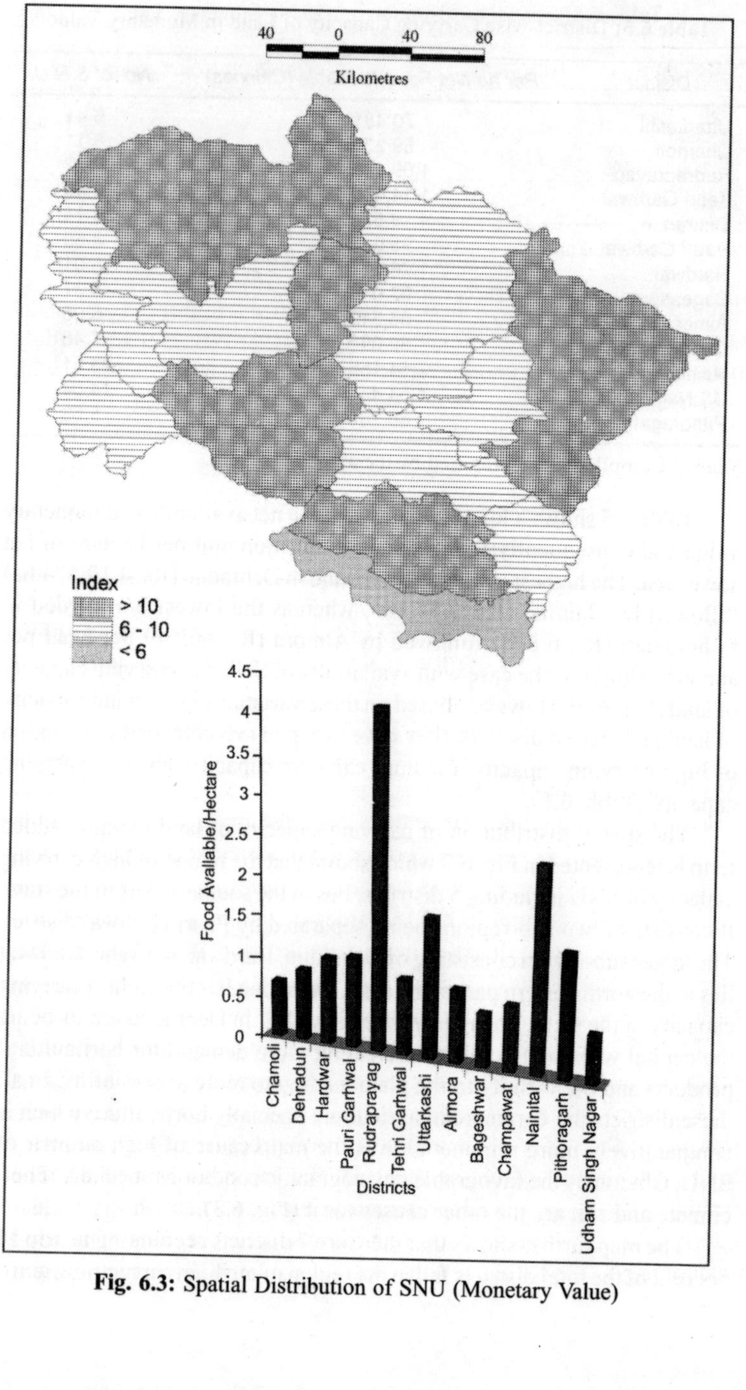

Fig. 6.3: Spatial Distribution of SNU (Monetary Value)

Table 6.6: Spatial Distribution of SNU in Monetary Value

Category	*SNU range*	*No. of districts*	*%age of total*
High	> 10	3	38.46
Medium	10-6	7	53.85
Low	< 6	1	7.69
Total		**13**	**100.0**

Source : Compiled and Computed by the author.

Except Pauri Garhwal, all these district lie in the northern part of the state including Greater Himalaya and substantial part of the middle Himalaya. Obviously non-conducive physical environment coupled with socio-institutional factors are the basic causes of comparatively low SNU. There is only one district of Pithoragarh which falls under low category. Obviously, the detrimental environmental conditions for agricultural activities are the basic reasons for it.

6.5.4 Spatial Pattern of Food Security

As has been pointed out earlier status food of security has been worked out in two terms i.e. (I) Calorific Value and (II) Monetary Value

6.5.4.1 Spatial Pattern of Food Security in Calorific Value

The spatial pattern of food security has been ascertained at district level on the basis of availability of calories per consumption unit. All edible agricultural products have been converted into calorific value for each district that gives the gross food available. It has been multiplied by the coefficient of production (0.832) in order to get the net available food for each district. Finally, the net available food in divided by the consumption units of each district and the 365 days of the year so as to get the amount of calories per consumption unit per day (Table 6.4).

Table 6.7 shows a lot of variations in net availability of calories. It ranges from a slow as 1985 calories in Pithoragarh district to as high as 3,255 calories in US Nagar district. Therefore, the districts have been arranged in ascending order and have been grouped into 5 categories based on the standard requirements of the calories (Table 6.8). The resulting pattern has been represented in Fig. 6.4 which shows a clear pattern of distribution of food security/insecurity regions. The food insecure region consisting of 6 districts substantially lies in the northern

Table 6.7: District-wise Consumption of Calories

Districts	*Conc. of Cal per Consumption Unit*
Chamoli	2,113
Dehradun	3,010
Hardwar	3,098
Pauri Garhwal	2,385
Rudraprayag	2,357
Tehri Garhwal	2,799
Uttarkashi	2,265
Almora	2,298
Bageshwar	2,725
Champawat	2,441
Nainital	2,913
Pithoragarh	1,985
US Nagar	3,255

Source: Compiled and Computed by the author.

part of the state. It covers the Greater Himalaya and northern part of the Middle Himalayas. Harsh climatic conditions coupled with adverse physiography with poor mountain soil are the basic causes of low agriculture production leading to food insecurity. Based on availability of calories this region has further been divided into two sub-regions viz. Severely insecure region and insecure region. The former consists of two districts of Chamoli and Pithoragarh, while the latter includes 4 districts of Uttarkashi, Rudraprayag, Pauri Garhwal and Almora. The over all physical conditions prevailing in these two sub-regions speak much about their differential level of food security.

The food secure region consists of 7 districts which are mainly in Tarai plains and southern part of the Middle Himalayas. Overall favourable physiographic conditions together with proximity to plains leading to more exposure to technological agricultural inputs are the main causes of food security in this region. But there are intra-regional variations in it also. Hence, it has been divided into 3 sub-regions based on availability of calories. The food secure region includes 2 districts of Bageshwar and Champawat. The lower valleys with fertile soils of the major river of these districts are the main causes of comparatively high agricultural production leading to more food security. The moderately food security region with calorific availability between 2,730—3,030 covers the districts of Dehradun, Hardwar and Nainital. Obviously, these districts have comparatively much better physiographic conditions and high level of

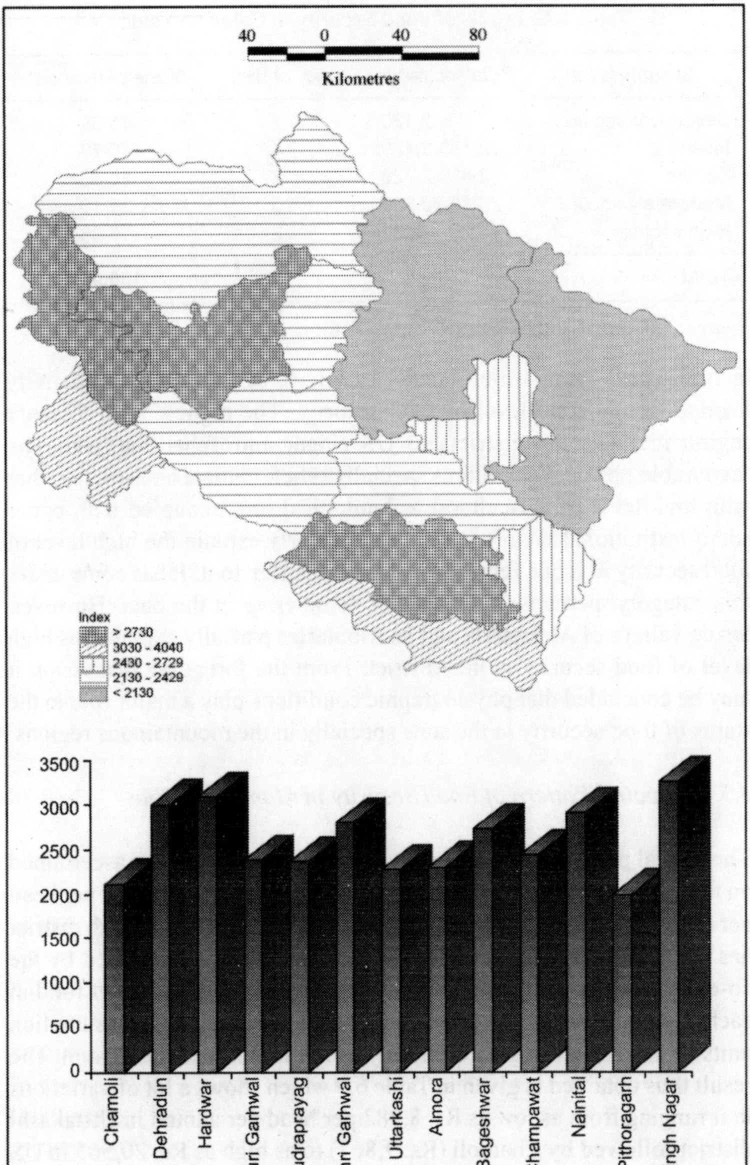

Fig. 6.4: Level of Food Security in Calorific Value

Table 6.8: Levels of Food Security in Calorific Value

Security level	*Caloric range*	*No. of dist.*	*%age of total dist.*
Severely insecure	< 2,130	2	15.38
Insecure	2,130-2,429	4	30.76
Secure	2,430-2,729	2	15.38
Moderately secure	2,730-30,380	3	23.10
High secure	>3030	2	15.38
Total		**13**	**100.00**

Source: Compiled and Computed by the author.

technological input which results in higher agricultural productivity leading to moderate level of food security. The highest food insecure region includes the districts of US Nagar and Tehri Garhwal. The favourable physical conditions specially relief, climate and soil together with high level of agricultural technological input coupled with better socio-institutional factors in US Nagar clearly explain the high level of food security in it but Tehri Garhwal is exception to it. It has come under this category, perhaps, by chance or some error in the data. However, fertile valleys of Alaknanda and its tributaries partially explain this high level of food security in this district. From the foregoing discussion it may be concluded that physiographic conditions play a major role in the status of food security in the state specially in the mountainous regions.

6.5.4.2 Spatial Pattern of Food Security in Monetary Value

The spatial pattern of food security at district level has been ascertained on the basis of availability of net food in value added terms and per head per day of consumption unit. All the agricultural produce of each district has been converted in monetary value and has been multiplied by the co-efficient of production in order to get the net availability of food in each district. It has been, thereafter, divided by the total consumption units of the district to get availability of funds per head per annum. The result thus obtained is given in Table 6.9 which shows a lot of variations in it ranging from as low as Rs. 8,982 per head per annum in Uttrakashi district followed by Chamoli (Rs. 9,867) to as high as Rs. 20,565 in US Nagar followed by Dehradun (Rs. 17,643) (Fig. 6.5). However, the results emerging from the Table 6.9 clearly show the strong positive relationship between the physiographic conditions and availability of funds.

Table 6.9: District-wise Availability of Monetary Value

District	*Cons. per conc. unit p.a.*
Chamoli	9,687
Dehradun	17,643
Hardwar	16,268
Pauri Garhwal	14,570
Rudraprayag	11,817
Tehri Garhwal	14,328
Uttarkashi	8,982
Almora	11,127
Bageshwar	12,387
Champawat	10,437
Pithoragarh	7,595
US Nagar	20,565

Source : Compiled and Computed by the author.

However, based on availability of funds per consumption unit per annum, the districts have been arranged in descending order and have been grouped into 5 food security regions i.e. severely insecure, insecure, secure, moderately secure and highly secure (Table 6.10). It has been shown on Fig.6.5. The food insecure region covering 4 districts lies in a contiguous belt in northern and eastern part of the state including Uttarkashi, Chamoli, Pithoragarh and Champawat districts. This region has further been divided into two sub-regions on the basis of availability of fund per consumption unit. Uttarkashi and Pithoragarh fall in the severely insecure region with less than Rs. 9,000 availability per consumption unit per annum, while Chamoli and Champawat come in

Table 6.10: Level of Food Security in Monetary Value

Security Level	*Monetary Range*	*No. of Distt.*	*%age of total Distt.*
Severely Insecure	< 9,000	2	15.38
Insecure	9,000-10,999	2	15.38
Secure	11,000-12,999	3	23.10
Moderately Secure	13,000-15,000	2	15.38
Highly Secure	>15,000	4	30.76
Total		**13**	**100.00**

Source: Compiled and Computed by the author

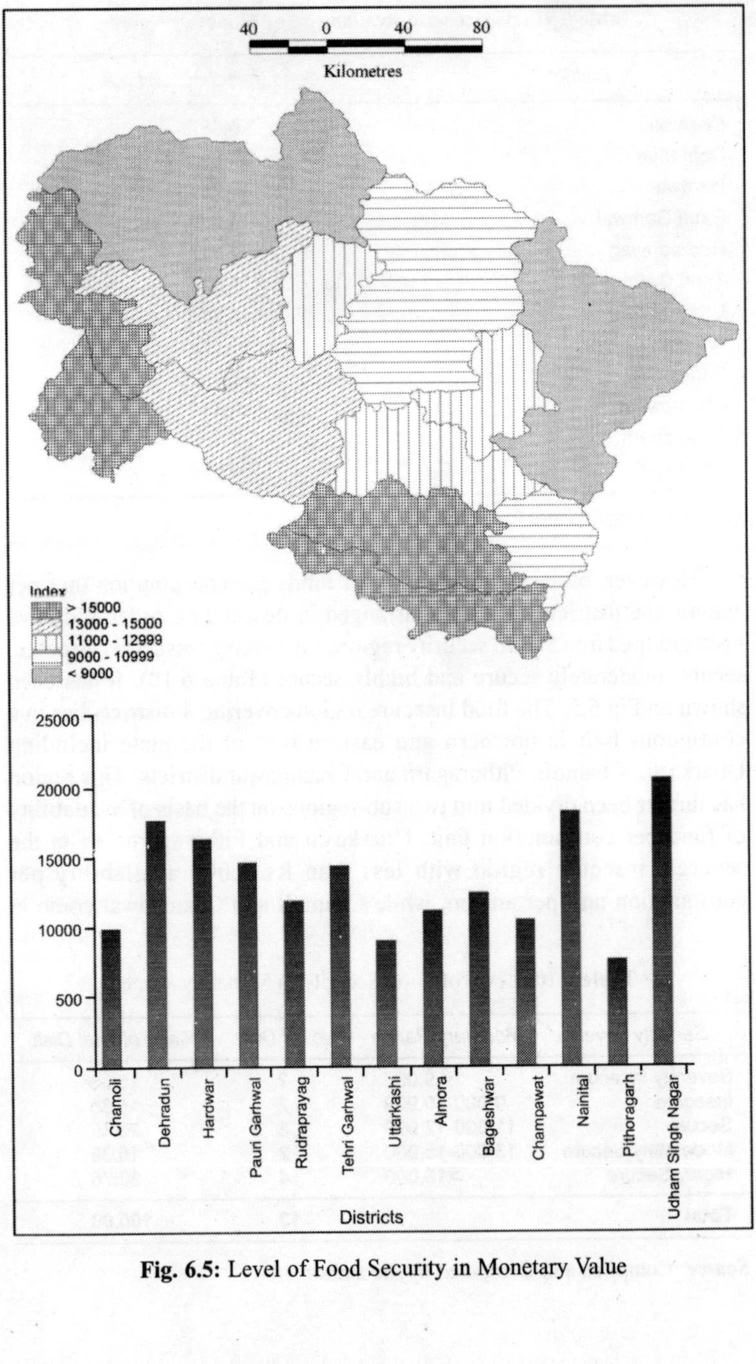

Fig. 6.5: Level of Food Security in Monetary Value

the food insecure region. A total of 9 districts covering nearly 73 per cent of the districts have food secure status. There are 3 district viz. Rudraprayag, Bageshwar and Almora which have the status of food security, while two districts of Pauri Garhwal and Tehri Garhwal have moderate level of food security. The largest number of districts i.e. 4 districts of Dehradun, Hardwar, Nainital and US Nagar constituting nearly 31 per cent of the total districts have high level of food security with more than Rs.15,000 per consumption unit per annum. The physiographic, techno-economic and socio-cultural factors speak a lot about their high level of food security. Besides this, commercial agriculture and enterprising farmers are other reasons for it.

From the foregoing analysis it becomes quite obvious that the level of food security increases from the northern mountainous regions towards the foothill plains.

References

Asian Development Bank, 2004 : *Agriculture, Food Security and Rural Development.*

Aykroyd, W.R., 1956 : *The Nutritive Value of Indian Diet and Planning of a Satisfactory Diet*, Madras.

Bader, M.D.M., 2010 : *Disparities in Neighbourhood Food Environment : Implications : Measurement Strategies*, Vol. 86(4).

Banerjee, A. (Ed.), 2004 : *Food Security and the Public Distribution System Today*, Kanishka Pub., New Delhi.

Barun, J. *et al.*, 1992 : *Improving Food Security of the Poor, Concept Policy and Programme*, International Food Policy Research Institute, Washington.

Bernard, F.E. and Derrick, J.T., 2007 : 'Population Pressure and Human Carrying Capacity in Selected Locations of Machakas and Ketni Districts', in Thakur, B. (Ed.), *Perspectives in Resource Management in Developing Countries.*

Bhattacharya, B., 2001 : Food Security, NAP and NATO, *Yojana*, Vol. 45, No.1

Borlang, N.E., 2004 : 'Feeding a World of 10 Billion People : Our 21st Century Challenge' in Scanes, C.G. and Miranowski, J.A. (Eds.), *Perspectives in World Food and Agriculture*, Blackwell Pub. Co. Iwa.

Chakravarty, A.K., 1970 : Food Sufficiency Patterns in India; *Geographical Review*, Vol. LX, New York.

Chung, K., *et al.*, 1997 : *Identifying the Food Insecure : The Application of*

Mixed Method Approaches in India; International Food Policy Research Institute, Washington.

Colin, C.G. and Miranowski, J.A., 2004 : *Perspectives in World Food and Agriculture*, Blackwell Pub. Co., Iwo.

Copra, *et al.*, (Eds.), 1998 : *Crop Productivity and Sustainability—Shaping the Future*, Oxford and IBH Pub. Co., New Delhi.

FAO 1987 : *The Status Food Insecurity in the World*, Rome.

FAO 1999 : *Food Outlook*, No.4, Rome.

Frankenberger, T.R. and McCarton, 1998 : The Household Livelihood Security Concept, *Food Nutrition and Agriculture*, FAO, Rome.

George, P.S., 1994 : Food Security in South Asia : Performance and Prospects, *Economic and Political Weekly*, Vol. 29, No. 15, April 30.

George, P.S., 1994 : Public Distribution System, Food Subsidy and Production Insecurities, *Economic and Political Weekly*, Vol.31, No.39, September 28.

George, P.S., 1999 : Some Reflection on Food Security in India, *Indian Journal of Agricultural Economics*, Vol. 54, No. 2.

Gopalan, C. *et al.*, 2000 : *Nutritive Value of Indian Foods*, Indian Council of Medical Research, Hyderabad.

Gopalan, C., 2000 : Medium Term Strategy for Meeting the Nutritional Needs of the Population, *Population and Human Development—Meeting Some Critical Needs in the New Century*, Commissioned Papers, Planning Commission.

Johnson, G., 1976 : Increased Stability of Grain Supplies in Developing countries, Optional Carry Overs and Insurance, *World Development*, Vol. 4, No. 12.

Kendall, M.G. 1939 : The Geographical Distribution of Crop Productivity in England, *Journal of Rural Statistical Society*, Vol. 38.

Khusro, A.M., 1965 : Measurement of Agricultural Productivity at Macro and Micro Level, *Journal of Indian Society of Agricultural Statistics*, Vol. 17(2).

Masood, S. (Ed.), 2009 : *Peri-Urban Agriculture : Agriculture by Twenty and Beyond*, A.P.H. Pub. Corp., New Delhi.

Minhas, B.S., 1976 : Towards National Food Security, *Indian Journal of Agricultural Economics*, Vol. 31, No. 4.

Mohammad, N. (Ed.), 1992 : *New Dimensions in Agricultural Geography*, Vol. I-VIII, Concept, New Delhi.

Mohammad, N. 2003 : 'Regional Pattern of Food Security in India' in Singh, T.P. (Ed.) *Resource Conservation and Food Security*, Concept. Pub. Co., New Delhi.

Murlidharan, S., 2001 : Upholding the Right to Food, *Frontline*, Vol.18(18), New Delhi.

Radhakrishna, R., 1991 : Food and Nutrition : Challenge for Policy, *Journal of the Indian Society of Agricultural Statistics*, Vol. 53, No. 3, December.

Raza, M. 1992 : Regional Dimensions of Agricultural Development in Mohammad, N. (ed.), *New Dimensions of Agriculture*, Vol.IV, Concept Pub. Co, New Delhi.

Reutlinger, S., 177 : *Food Insecurity : Magnitudes and Remedies*, Washington, U.S.A.

Sarris, A. and Taylor, L., 1976 : Central Stock, Food Aid and Food Security, Food, *Nutrition and Agriculture*, Vol. 4, No. 2.

Sen, A.K., 1981 : *Poverty and Famines : An Essay in Entitlement and Deprivation*, Clarendon Press, Oxford.

Shafi, M. and Aziz, A. (Ed.), 1989 : *Food Systems of the World*, Rawat Pub., Jaipur.

Shafi, M., 1972 : Measurement of Agricultural Productivity of the Great Indian Plains, *The Geographer*, Vol. 19(1), Aligarh.

Singh, B., 1948 : *Population and Food Planning in India*, Bombay.

Singh, J. 1972 : A New Technique of Measuring Agricultural Efficiency in Haryana, India, *The Geographer*, Vol.19(1) Aligarh.

Singh, S. 1948 : *Concept of Computation of Consumption Unit*.

Singh, S. 2001 : Human Development and Food Security, *Yojana*, Vol. 45, No. 8.

Singh, V.R., 1970 : Land Capability—A Method of Classification, *Professional Geographer*, Vol. 22(1).

Stamp, L.D., 1958 : The Measurement of Land Resources, *The Geographical Review*, Vol. 48(1).

Suryanarayana, M.H., 1996 : Food Security and Calorie Adequacy Across States, *Journal of Indian School of Political Economy*, Vol. 7, No. 2.

Swaminathan, M.S. 1993 : Food Security through Sustainable Agriculture, in *Biotechnological Applications for Food Security in Developing Countries*, Oxford and IBH Pub. Co., New Delhi.

Swaminathan, M.S. 2001 : Food Security Atlas of Rural India, *Frontline*, Vol. 18.

Valdes, A. ed.), *Food Security for Developing Countries*, Westview Press, Colorado.

World Bank Development Report, 1986.

Chapter 7

Agricultural Diversification and Food Security

A Microlevel Analysis

7.1 Introduction

Agriculture is not only the product of physical environment but also a set of techno-economic and socio-cultural framework such as adoption of agricultural innovations, land ownership, size, shape and layout of farms, educational level, age structure, agricultural implements, market centres, etc. There is no general theory of socio-eco-cultural composition which could be fairly applied to explain the spatial variations in agricultural system but social and economic differences within communities and between communities do affect the scale and type of farming operations together with choice of enterprises. The decision of the farmers to put the land to different uses depends upon their age, education, sex, occupation, psychology and so on (Bose, 1961; Lee, 1960 and Mohammad, 1988). In this chapter, therefore, an attempt has been made to discuss these and other personal, household and farming characteristics of the households to assess the status of agricultural diversification and food security.

7.2 Personal and Household Characteristics

The personal and household characteristics of the households and their families have a direct bearing on primary occupation specially agriculture, exposure to impersonal communication, adoption of agricultural innovations, level of knowledge and involvement etc. Therefore, some of the important socio-cultural components have been analysed here.

7.2.1 Age Composition of Households

Adoption of new idea, practices and technology helps to develop the agricultural system of an area. In this process age composition plays an important role. The presumption is that at a comparatively younger age people are more receptive to new ideas and practice, whereas, at the old age they find it difficult to change from age old practices and they resist to adoption of innovations (Mohammad, 1988). With this view all the 538 households of the 30 selected villages have been arranged in ascending order and have been grouped into 4 categories (Table 7.1).

Table 7.1: Age Composition of Households

Age Group	*Age range*	*No. of respondents*	*% to total*
Young Age	(<30) years	61	11.34
Middle Age	(31-45) years	207	38.48
Upper Middle Age	(46-59) years	132	24.54
Old Age	(>60) years	138	25.65
Total		**538**	**100.00**

Source: Compiled and Computed by the author.

Table 7.1 shows that there are 61 households accounting a little more than one-tenth of the total households are in the young age group of less than 30 years. The largest number of households i.e. 207 amounting to nearly 38 per cent of the total are in the middle age-group of between 31 and 45 years. The old age households of more than 60 years account nearly one-fourth of the households, while the upper middle age group households (132) account nearly 24.5 per cent of the total households (Fig. 7.1). The above analysis clearly indicates that the younger population, who is comparatively more educated and innovative, either takes little interest in agriculture and is interested in white collar jobs or is being given little freedom to agricultural operations. The middle age households within the age-group of 31 to 60 amounting to 63 per cent of the households take more interest in agricultural activities. Interestingly nearly one-fourth of households belonging to old age-group who take much interest in agricultural activities. It shows that social values and norms dominate the agricultural activities. By and large, father or elder brother, by virtue of their being older in age, is not only the leader of the family but his decision is final in day to day routines including agricultural activities.

The above analysis represents the aggregate picture of the age composition of the households but there are a lot of inter and intra-village

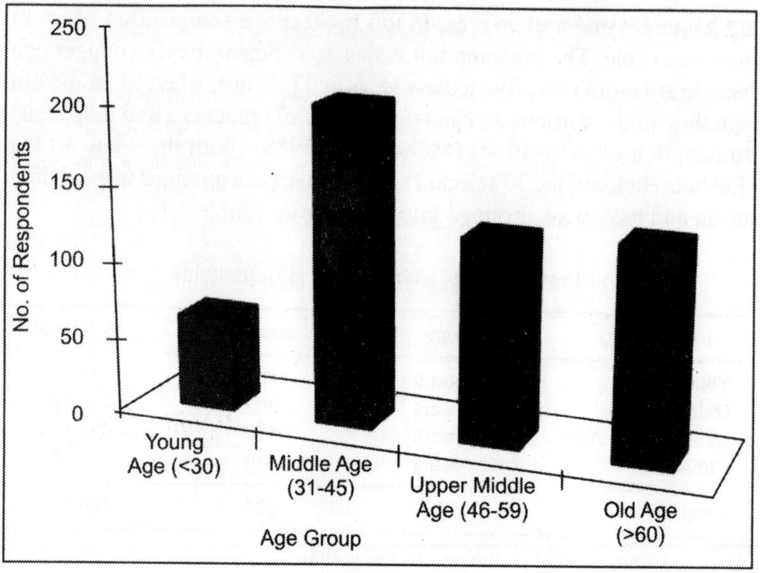

Fig. 7.1: Age Comparison of Respondents

variations in it (Appendix-XI). There are 10 villages where young age households are absent, while in 9 villages, they account less than 10 per cent of the households. In 6 villages, the young households account 10-20 per cent, while in remaining 5 villages they account more than 20 per cent of the households. The distributional pattern of middle and upper middle age households is better. In 50 per cent villages, more than 40 per cent households belong to middle age-group, while in 12 villages 20 to 40 per cent households belong to this category and in remaining 3 villages less than 20 per cent households fall in this group.

Although the distributional pattern of households is heterogeneous but, by and large, in the villages of Greater Himalaya and northern part of Middle Himalaya, the old age households contribute the larger percentage to the total households, while just opposite is the case in the villages of the southern Middle Himalaya and the Siwalik and foothills. The basic reason, perhaps, is that these villages are much exposed to the plain and its innovative people. Moreover, the agriculture is more remunerative here because of fertile soil and available infrastructural facilities.

7.2.2 Age Structure of Household Members

The age structure of the household members gives an idea about the nature and amount of dependency and the workforce. In certain cases it is a decisive force in selection of occupation and crop diversification. With this view the household members of all the selected households of the 30 selected villages have been classified into four groups (Table 7.2).

Table 7.2: Age Composition of household members

Age Group	*Age range*	*Persons*	*% age to total*
Children	< 14 years	524	19.29
Young age	14-30 years	321	11.82
Middle age	31-45 years	859	31.62
Upper middle age	46-60 years	619	22.80
Old age	>60 years	393	14.47
Total		**2716**	**100.00**

Source: Compiled and Computed by the author.

Table 7.2 indicates that nearly one-fifth of the household members are children of less than 14 years of age, while about 15 per cent belong to old age-group of more than 60 years of age which shows that nearly one-third of the household members are, by and large, dependent on the rest of the population. And, it shows a high level of dependency. The largest number amounting to nearly 32 per cent of the total members are in the middle age-group of 31 to 45 years. It is followed by the upper middle age-group which accounts nearly 23 per cent of the household members. As such nearly 64 per cent of the household members are in the age-group of 31 to 60 years, while young age-group accounts only 12 per cent of the total population. On the whole, it may be concluded that over all dependency ratio is not going to reduce in near future (Fig.7.2).

Like the age structures of the respondent, there are a lot of inter and intra-village variations in it. By and large, the dependency ratio is less in the southwestern part of the state and it goes on increasing towards its northeastern part.

7.2.3 Educational Status of Households

The role of education in the development of agriculture hardly needs any

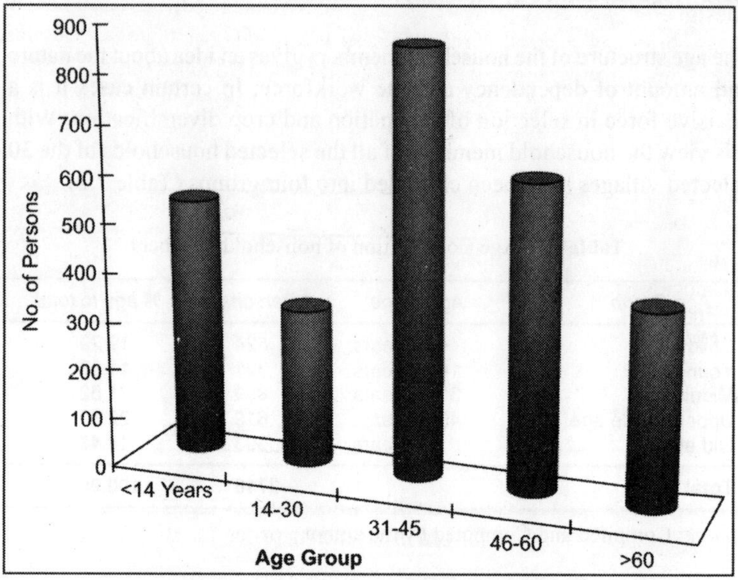

Fig. 7.2: Age Comparison of Household Members

emphasis. It is one of the chief instruments through which a society socializes its members and brings desirable changes in the social life of its people. In fact education is the aggregate of all the processes by means of which a person develops abilities, attitudes, and other forms of behaviour of positive values in the society he lives or it is a social process by which people are subjected to the influence of selected and controlled environment so that they can attain social competence and optimum individual development (Bear, 1947). And as such it could be argued that education influences the course of behaviour of both society and its individuals. And adoption of farming occupation and related activities are not exception to it. Various studies have proved that there is a close relationship between educational level and adoption of agricultural innovations which in turn affects agricultural diversification (Mohammad, 1981 and Rangaswamy, 1972).

However, in order to ascertain the educational status of the individual households, questions on formal and informal education of households were included in the questionnaire. Different answers varying from being illiterate to Ph.D. were received. So it was difficult to categories the households into any meaningful category on this basis, howsoever well designed it may be. Therefore, in order to assess the educational status of households at uniform level, and also to classify them into reasonable categories, following scoring scheme has been adopted (Verma, 2007). This scheme may be subjective but utmost care has been taken to maintain objectivity.

Literacy Standard	*No. of Scores*
Illiterate	0.00
Informal Education	0.25
Primary School	0.50
Middle School	0.75
High School	1.25
Senior Secondary	1.75
Graduation	2.50
M.A. & Others	5.00

After assigning the scores, the total scores of every individual was found out which were arranged in ascending order. Finally, the households were divided in 5 categories as given in Table 7.2.

Table 7.3: Educational Status of Households

Groups	*Score range*	*No. of Respondents*	*% to total*
Illiterate	– 0.00	108	20.07
Less Edu.	0.00 – 1.50	276	51.30
Moderately Edu.	1.50 – 2.50	98	18.22
Highly Edu.	> 2.50	56	10.41
Total		**538**	**100.00**

Source: Compiled and Computed by the author.

Table 7.3 indicates that more than one-fifth of the households are illiterate, while more than half of the households are less educated with less than 1.50 educational score. There an 98 households accounting about 18 per cent of the total who are moderately educated, while nearly one-tenth households belong to highly educated category with more than 2.5 educational score. It may be inferred from the above tabular analysis that the literacy level among the households is low. Only one-fourth of the households are reasonably educated.

The above analysis represents the aggregate picture of the educational status of the households but there are a lot of inter and intra-village

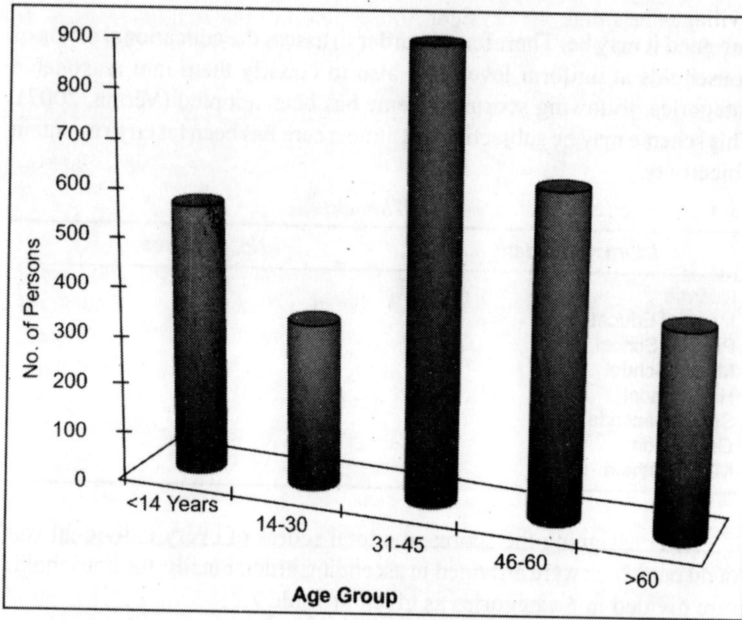

Fig. 7.3: Educational Status of Respondents

variations in the educational level of the households (Appendix-XII). There are 11 villages where more than 20 per cent households are illiterate, and in equal number of villages 10 to 20 per cent households are illiterate, while in remaining 8 village, the percentage of illiterate households is less than 10. The status in case of highly educated households with more than 2.50 educational score is just the opposite of it. There are 3 villages where more than 20 per cent households are highly educated, while there are 18 villages including 5 villages with none of the households in this category, where less than 10 per cent households are under highly educated category. In remaining 9 villages 10 to 20 per cent households fall under this category. In the moderately educated category with 1.50 to 2.50 educational score, the distributional pattern is more or less uniform. There are 12 villages where 20 per cent households fall under this category, while 9 villages have 0-20 per cent households under this category. In 3 villages there is none in this category, while in 6 villages less than 10 per cent households fall under this category. In the less educated category the distributional pattern is quite uneven. There an 14 villages where 40-60 per cent households fall under this category, while in 9 villages more than 60 per cent households are under this group. In the remaining

7 villages less than 40 per cent households fall under this category. It may be generalized from the above discussion that, by and large, in the foothill plains and tarai region, the educational level is high, while in the high mountainous areas of the northeastern part of the state, the educational level is low.

7.2.4 Educational Index of the Households

Education plays an important role in the decision-making process. In the overall process of agricultural development and agricultural diversification, it is not only the educational status of the households, that matters but also the educational level of all the family members. Educated households easily accept innovations in every field as compared to less educated households. The surveyed household members have educational status ranging from informal education to post-graduation. Hence, it is not possible to aggregate the educational status of the household due to varied size of family and different educational status. Therefore, the 'Educational Index of the Households' has been worked out with the help of scoring scheme given earlier. After assigning scores to the family members, their scores were added together and divided by total member of the household excluding children below 5 years of age. The results thus obtained have been arranged in ascending order and the households have been divided in 5 categories (Table 7.4).

Table 7.4 indicates that more than one-fifth (21.98%) household do not have even a single members in the family who is educated, while nearly 50 per cent household fall under less educated category. Thus nearly 70 per cent household members are very poorly educated, while 486 households members amounting to nearly 20 per cent households are moderately educated. Only 222 household members accounting 9 per cent of the households fall under highly educated category.

Table 7.4: Educational Index of Household

Groups	*Score range*	*No. of person*	*%age to total*
Illiterate	0 –0.00	538	21.98
Less Educated	0.00 – 1.50	1201	49.07
Moderately Educated	1.51 – 2.50	486	19.86
Highly Educated	>2.50	222	9.09
Total		**2447**	**100.00**

Source: Compiled and Computed by the author.

A close look at the household educational level of the households clearly indicates that in the villages located in the vicinity of plains or major towns like Nainital or Dehradun, the educational level is high due to availability of educational infrastructure, while in the remaining areas the educational level is low.

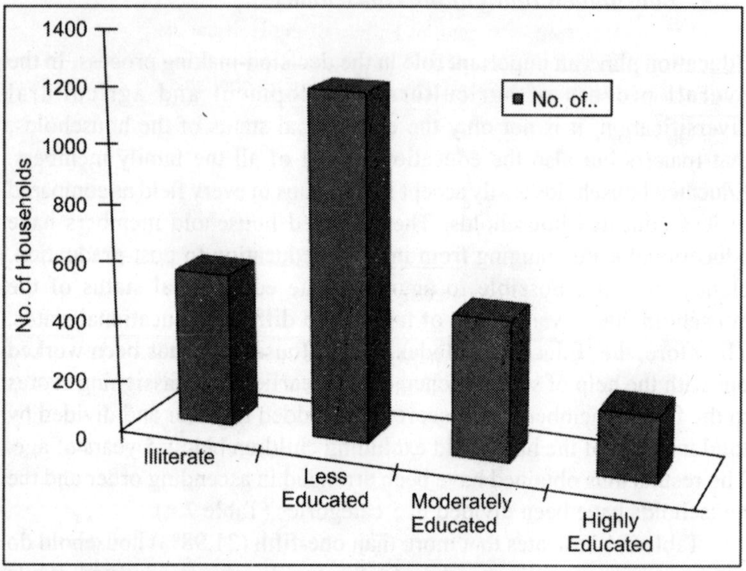

Fig. 7.4: Composite Index of Education of Household Members

7.2.5 Family Size of Households

The old agrarian societies of the orient are characterized as familistis, and the whole social organisation and the values and beliefs of the people are centered around social institutions. Most of the families, specially in the countryside, are large in size and extended or joint in composition. Parents generally live with their married sons and their families. By and large, it is the family, instead of the individuals, which is the unit of social action. The thought and actions of the individual members in the family, in a large measure, are influenced by the family as a single entity (Mohammad, 1988). The personality of a member of a large family is somewhat different from that of a member belonging to a small family. Hence, the size of a family has a major role in deciding the type of occupation to be adopted as well as in the process of its development. Therefore, the total members of the selected households were enumerated during the course of the field work. In the present study the family includes all the members of selected households who live in one household and share a common kitchen. On the basis of the total members of the household, households were arranged in ascending order and finally grouped into five categories as given in Table 7.5.

Table 7.5: Size of Family of the Households

Family Size	*Members range*	*No. of respondents*	*% to Total*
Small	<4	112	20.82
Medium	5-9	335	62.27
Large	10-14	62	11.52
Very Large	>14	29	5.39
Total		**538**	**100.00**

Source: Compiled and Computed by the author.

Table 7.5 shows that on the whole the size of the family of households is reasonable. There are 112 households amounting to nearly one-fifth of the total households who have small size of family with less than 4 persons. Substantial number of households i.e. 335 persons accounting to 62 per

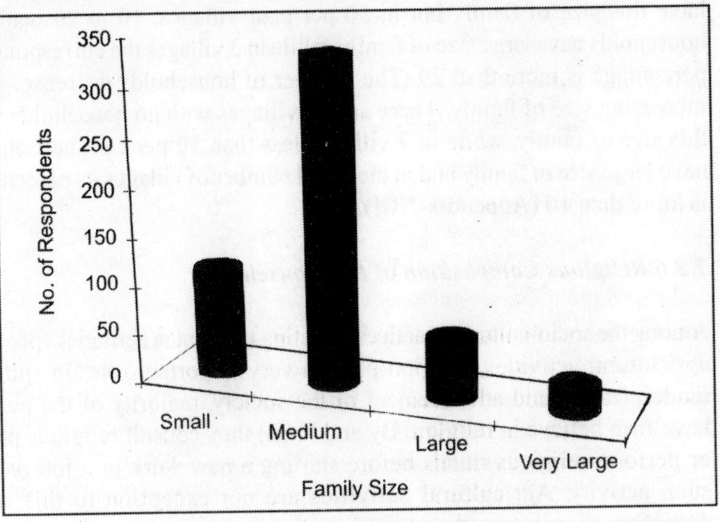

Fig. 7.5: Size of Family of Respondents

cent of the total households have a medium size of family with 4 to 9 member. Most of these households belong to the northwestern mountainous region of the state where the means of sustenance is poor and they are leading a harsh life. Slightly more than one-tenth of the households have a large size of family, while only 29 households accounting 5 per cent of the households have very large size of family with more than 10 member in it. Most of these households belong to the foothill plains and tarai regions where there is better availability of means of sustenance (Fig. 7.5).

The foregoing analysis presents an aggregate picture of the size of the family of households but there are a lot of inter and intra-village variations in it (Appendix-XIII). In 16 villages, less than 20 per cent households have small size of family, while in 10 villages, 20 to 30 per cent households have this size of family. In remaining 4 villages more than 30 per cent households fall under this category.

In case of medium size of family, the picture is quite different. In 6 villages less than 50 per cent households have medium size of family, while in 16 village, 50 to 70 per cent households have this size of family. In remaining 8 villages, more than 70 per cent households belong to this category. The number of households with large size of family is small (62) and in 5 villages there is not even a single respondent in this category. However, there are 7 villages where less than 10 per cent households have this size of family but in 50 per cent villages, 10 to 20 per cent households have large size of family, while in 3 villages the corresponding percentage is more than 20. The number of households decreases with increasing size of family. There are 16 villages with no households with this size of family, while in 7 villages less than 10 per cent households have large size of family and in the equal number of villages its percentage is more than 10 (Appendix-XIII).

7.2.6 Religious Composition of the Households

Among the socio-cultural practices affecting the human activities specially agricultural activities, religion plays a very important role. In spite of modernization and advancement of the society, majority of the people have firm believe in religion. By and large, they consult religious priest or perform religious rituals before starting a new work or a job or any such activity. Agricultural activities are not exception to this rule. Therefore, there is a need to study the religious composition of households which is given in Table 7.6 Fig. 7.6..

Table 7.6: Religious Composition of Households

Religious Group	*No. of Husehold*s	*% to Total*
Hindu	504	93.68
Muslim	92	4.09
Sikh	12	2.23
Total	**538**	**100.00**

Table 7.6 clearly indicates that the region is predominantly inhabited by Hindus. There are 304 households amounting to nearly 94 per cent of the total households who belong to Hindu community. Only 22 households covering 4 per cent of the total households are Muslims, while just 12 households covering only 2 per cent of the households

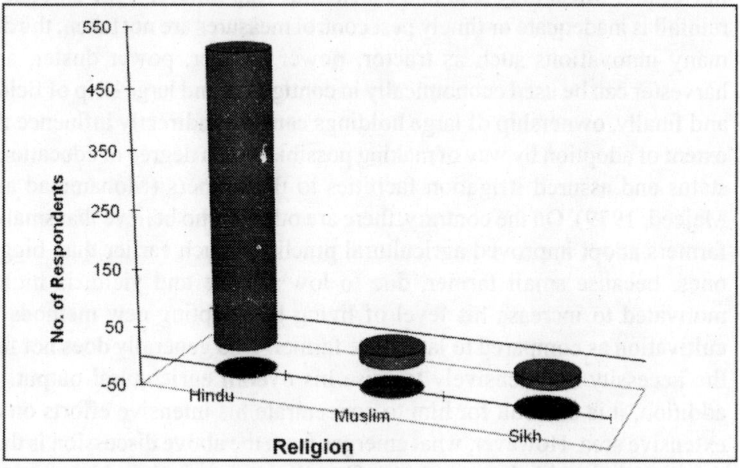

Fig. 7.6: Religious Composition of Households

belong to Sikh religion. The Muslims are mostly found in Hardwar and Dehradun districts, while the Sikhs are mainly concentrated in US Nagar. Out of 22 Muslims 21 belong to Rasoolpur village of Hardwar and only one households belongs to Malsa village, while all the 12 Sikh households belong to US Nagar.

7.3 Farming Characteristic of Households

7.3.1 Size of Holding

The sociological and socio-anthropological studies provide enough evidences to prove the fact that the extent of ownership of land is an important indicator of one's socio-economic status (Sinha, 2009). It is more so in an essentially agricultural country like India, where land, being scarce, is much valued by the people. Whatever, may be the case, there are two contrary views regarding the size of holding and agricultural development and its diversification which is, in fact, a result of adoption of agricultural innovations. There is a positive and high correlation between the extent of land ownership and adoption of agricultural innovation by the farmers due to many reasons : Firstly, many innovations such as new agricultural implements—tractor, pumping set, thresher, etc, require substantial capital outlay which is beyond capital resources of smaller cultivators; secondly, the adoption of innovations is a matter of risk because sometimes a thoroughly tested innovation like many varieties of rice and wheat may lead to a substantial crop failure, if the amount of rainfall is inadequate or timely pest control measures are not taken; thirdly, many innovations such as tractor, power sprayer, power duster, and harvester can be used economically in contiguous and large strip of fields; and finally, ownership of large holdings can also indirectly influence the extent of adoption by way of making possible a high degree of educational status and assured irrigation facilities to the farmers (Mohammad and Majeed, 1979). On the contrary, there are others, who believe that smaller farmers adopt improved agricultural practices much earlier than bigger ones, because small farmer, due to low income and yield, is highly motivated to increase his level of living by adopting new methods of cultivation as compared to large size farmer, who generally does not feel the necessity so intensively to raise his overall agricultural output. In addition, it is difficult for him to concentrate his intensive efforts on an extensive area. However, what emerges from the above discussion is that, there is relationship between size of holding and agricultural innovations which in turn affects agricultural development and its diversification. In the present study attempt has been made to analyse two types of holdings viz. actual size of holdings and operational size of holdings.

7.3.1.1 Actual Size of Holdings

The actual size of holding includes all the land entered in the name of an individual in government records irrespective of the fact whether he cultivates it or not. Even the area under groves, pastures, waste lands, etc. are also included in this category, if it is in the name of an individual. But in the present study only agricultural land owned by an individual has been considered, and on this basis, the selected households have been classified as given in Table 7.7.

Table 7.7: Actual Size of Holding

Size of holding	*Size range*	*No. of Households*	*% to Total*
Landless	0.00	7	1.3
Marginal	< 1.00	367	68.22
Small	1.00-2.00	67	12.45
Semi-Medium	2.00-4.00	58	10.78
Medium	> 4.00	39	7.25
Total		**538**	**100.00**

Source: Compiled and Computed by the author.

Table 7.7 clearly reveals the status of poverty of the households. The size of holdings of the households ranges from zero hectare to about 5 hectares but the substantial number of respondent i.e. 367 covering 68 per cent of the total households are marginal farmers with less than one hectare of land, while 7 households are landless and they work either as freelance labourer or as agricultural labourer.

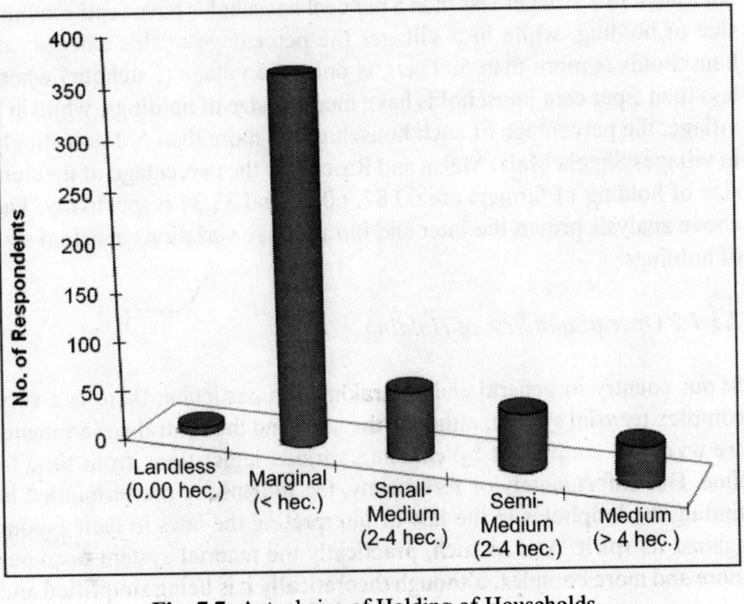

Fig. 7.7: Actual size of Holding of Households

More than 12 per cent (67 households) households have small size of holding with 1-2 hectares, while a little more than 10 per cent households (58) belong to the semi-medium size of holding. The remaining 39 households covering 7 per cent have medium size of holding with more than 4 hectares in size. None of the households have even large size of holding what to say of very large size of holdings.(Fig.7.7)

The above analysis of actual size of holding presents an aggregate picture but there are a lot of inter and intra-village variation in it (Appendix-XIV). There are only 4 villages i.e. Malsa (7), Syunsi (3), Rasoolpur (2) and Matipur (1) where landless households are found but in 18 villages more than 75 per cent households are marginal farmer, and in 5 villages 50-75 per cent households belong to marginal category, while in 7 villages less than 50 per cent households belong to this category. The small size of farmers are not found in 9 villages, while in equal number of villages less than 10 per cent households belong to this category. Only in 3 villages 10-20 per cent households fall under this category, while in 9 villages specially in Rasoolpur, Guleth and Maletha, more than 20 per cent households have small size of holdings. The situation in case of semi-medium and medium size of holding is very deplorable. There are 18 villages where semi-medium size of holdings of households are non-existent, while in 20 villages medium size of holding of households are not found. In 7 villages less than 5 per cent households have semi-medium size of holding, while in 5 villages the percentage of this category of households is more than 5. There is only one village (Fatehpur) where less than 5 per cent households have medium size of holdings, while in 8 village, the percentage of such households is more than 5. Interestingly in villages Sheela Mala, Malsa and Rasoolpur the percentage of medium size of holding of farmers are 60.87, 60.00 and 33.34 respectively. The above analysis proves the inter and intra-village variations in actual size of holdings.

7.3.1.2 Operational Size of Holding

In our country in general and Uttarakhand in particular, there is a very complex tenurial system, although the state and the central governments are trying to simplify it by enacting various legislations from time to time. But unfortunately or fortunately, the Indians are masterminded in finding the loopholes in the law or interpreting the laws in their favour against its spirit. And as such, practically the tenurial system becomes more and more complex, although theoretically it is being simplified and reformed. However, the owners of land, who are not in a position to cultivate their land themselves, or who do not want to cultivate the land, give their land to actual tiller on various terms and conditions. The nature of the term and conditions depends on the supply and demand of land to be leased out. If the demand is more and supply is meager, the owner of land exploits the tiller by giving his land to him (tiller) according to his term and conditions. On the contrary, when supply of land is more and demand is less, the tillers of land dictate the terms, and the owner has to accept it.

Nevertheless, whatever may be the tenancy system, it is the total area being cultivated by the households which matters more than the land registered in his name because it is in this size of holding in which he has to toil and supply all his inputs. Therefore, in the present study an attempt has been made to find out the operational size of holding of each households. After ascertaining the operational size of holding of each respondent, they have been arranged is ascending order and finally classified into 5 categories (Table 7.8).

Table 7.8: Operational Size of Holding

Size of Holding	*Size range*	*No. of Respondents*	*% to total*
Marginal	< 1.00 ha	378	70.26
Small	1.0-2.0 ha	61	11.33
Semi-Medium	2.0-4.0 ha	54	10.04
Medium	> 4.00 ha	45	8.37
Total		**538**	**100.00**

Source: Compiled and Computed by the author.

Table 7.8 represents almost the similar distributional pattern of the households. Nearly 70 per cent households (378 households) are marginal farmers with less than one hectare of land. The seven landless households have also taken the land from other persons for agricultural purposes. There are 61 households accounting nearly 11 per cent of the total with small size of holdings, while 54 households contributing nearly one-tenth of the total have semi-medium size of holding. The remaining 8 per cent have medium size of holding of more than 4 hectares.

The above analysis indicates that there is not much difference in the distributional pattern of the actual and operational size of holdings which shows that there is little system of giving the land on share cropping basis.

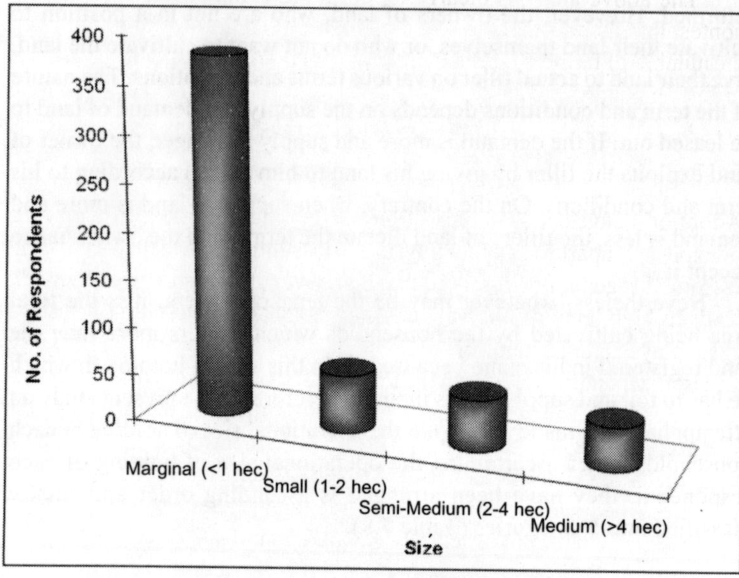

Fig. 7.8: Operational size of Holding of Respondents

The preceding paras give the aggregate picture of the distributional pattern of the operational size of holdings but there are a lot of inter and intra-village variations in it (Appendix-XV). There are only two villages (Rasoolpur and Sheela Mala) where marginal farmers are absent, while in 6 villages marginal farmers constitute less than 50 per cent of the households. But in 18 villages, marginal farmers share more than 75 per cent of the households. In the remaining 4 villages, the percentage of such farmers is between 50 and 75.

The households of small size of holding are not found in 11 villages, while in 7 villages, they account less than 10 per cent. In the remaining 12 villages, the households of small size of holdings account more than 10 per cent of the total. In case of the households with semi-medium size of holdings, there are 18 villages where they are non-existent, while in 3 villages their percentage is less than 10. In the remaining 9 villages they constitute more than 10 per cent of the total. The number of households goes on decreasing as the size of holding increases. There are 21 villages (70%) where households with medium size of holding have not been found, while in 4 village their percentage is less than 10. In the rest of the villages (5 Villages) they constitute more than 10 per cent.

The above analysis clearly supports the common view in national contest that there is inverse relationship between the size of holding and the number of farmers i.e. smaller the size of holdings larger is the number of farmers and *vice-versa*.

7.3.2 Level of Innovativeness

Adoption of agricultural innovations is very much associated with level of exposure, apart from other physio-socio-cultural and techno-economic factors. It is generally observed that high level of exposure to mass media leads to high level of innovativeness. Among the socio-cultural factors, caste plays an important role (Mohammad, 1981; Ramachandran, 1969; and Ramachandran, 1992). In the present study an attempt has been made to assess the level of innovativeness among different households. For this purposes various questions relating to major agricultural innovations viz. agricultural implements, improved varieties of seed, irrigation facilities, chemical fertilizers, insecticides and pesticides, were included in the questionnaire. In order to bring all the innovations and the several variations in their adoption at uniform level, a well thought of scoring scheme has been used (given in Chapter-2). The scores of individual households are added together in order to get his total score. The households are arranged in ascending order on the basis of their scores. Finally they have been grouped into 4 categories (Table 7.9).

Table 7.9: Level of innovativeness of Households

Category	*Score range*	*No. of respondents*	*% to total*
Least	< 4.0	494	91.82
Moderately	4.0-8.0	44	8.18
Highly	> 8.0	0	0.00
Total		**538**	**100.00**

Source: Compiled and Computed by the author.

The distribution of households is quite uneven among the various levels of innovativeness due to variations in their socio-economic conditions. Substantial percentage of households amounting to nearly 92 per cent of total households are least innovative, while a small percentage i.e., 8 is less innovative. There is none in highly innovative category. If we look at the spatial distribution of these households, they are concentrated either around the state and district headquarters or around the other urban centres. There are considerable variations in the level of innovativeness among the selected villages. There are 24 villages (75%) where all the households are least innovative. In the remaining 6 villages the households are either least innovative or moderately innovative (Appendix-XVI).

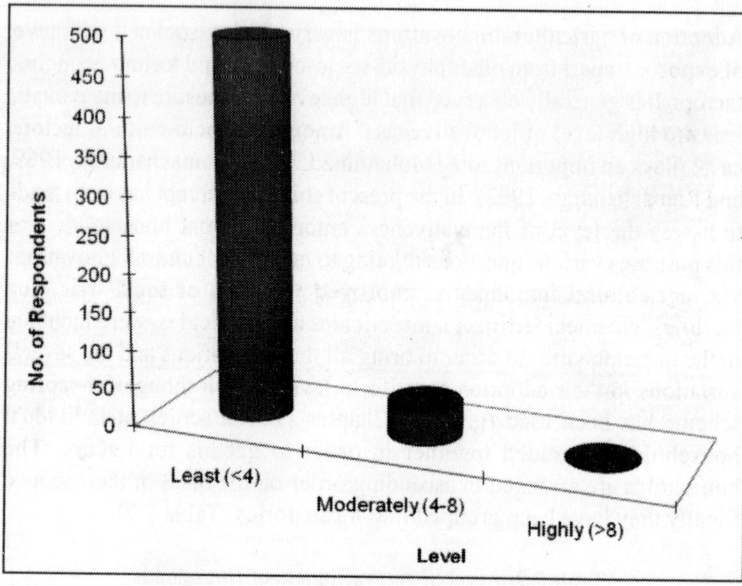

Fig. 7.9: Level of Innovativeness of Households

7.3.3 Spatial Patterns of Agricultural Landuse

No other economic activity is so dependent on environmental factors as agriculture. The physical factors particularly physiography, climate, drainage, soil and vegetation play very important role in shaping the agriculture of a region. Therefore, it is hardly surprising to note that there is variety in agricultural practices even over a small distance in Uttarakhand. These physical factors, which although do not now act in a deterministic fashion, lay down broad controls over what can be grown, and the yield which can be expected for a given input of land, labour, capital and other factors of production. Similarly, the role of socio-cultural factors hardly needs any emphasis. The interaction between the physical factors and socio-economic structure results in a great variety in agriculture. Such patterns become even more complex when we consider the farmers personal characteristics particularly his knowledge of new development in farming, and agriculture as a dynamic process. However, a brief account of the spatial patterns of agricultural land use among the selected households is given here.

Photo 7.1: Terrace Farming in Middle Himalaya

Like the great plains of India, in Uttarakhand also the agricultural year cannot be very clearly divided into three agricultural seasons like kharif, rabi and zaid. But for the purpose of convenience and more clear description, agricultural year is divided into two main seasons viz. kharif and rabi with a third season in between. The major crops of kharif season are rice, madua, jhingora, rajma, pulses, maize and sugarcane, while wheat, potato, barley, peas beans and oilseeds are the main crops of the rabi season. A brief account of cropping pattern of each season is given here.

7.3.3.1 Cropping Pattern in Kharif Season

Kharif season, by and large, sets in after the onset of the monsoon. It is an important agricultural season of the state. A statistical account of the area and percentage under each kharif crop is given in Table 7.10. Kharif crops occupy nearly 64 per cent of the gross cropped area of the households agricultural land. Rice, sugarcane, madua, jhingora and rajma are the major crops grown by the households during this season. Crop-wise brief account of main crops is given here.

Table 7.10: Village-wise Patterns of Land Use in Kharif Season

Crops	Area (Hec.)	% to total area
1. Rice	437.78	61.16
2. Mandua	37.87	5.29
3. Jhingora	2.90	0.41
4. Rajma	6.22	0.87
5. Pulses	3.38	0.47
6. Sugarcane	221.11	30.89
7. Other crops	6.51	0.91
Total	**715.77**	**100.00**

Source: Compiled and Computed by the author.

Rice covering 437 ha is the stepple crop of the households. It covers nearly 61 per cent of the total cropped area in this season or about 39 per cent of the gross cropped area of the respondent's agricultural land. It is widely grown by all the households across the state except in Bargaon

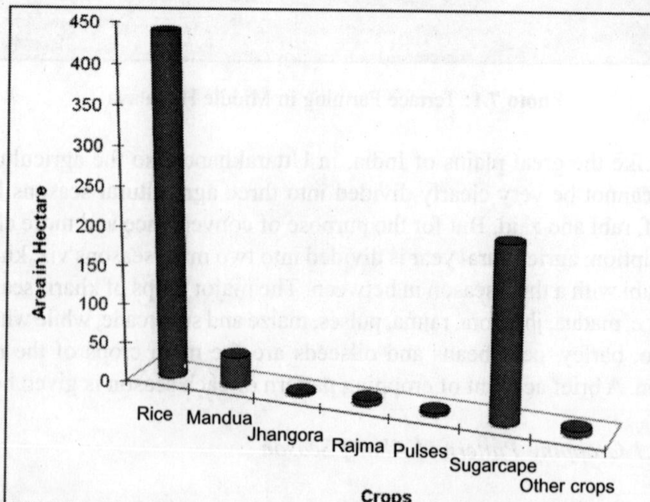

Fig. 7.10: Village-wise Pattern of Landuse in Kharif Season : 2005-06

where physical conditions are not conducive for its cultivation. But its spatial distribution is quite uneven (Appendix-XVII). There are 18 villages where rice occupies less than 5 ha of land, while in 5 villages, its areal strength is between 5 and 10 hectares. In the remaining 6 villages it occupies more than 10 ha with the exception of villages Malsa, Shahpur, Deonagar and Rasoolpur where it occupies 181.22, 74.8, 26.7, 24.88 ha respectively. By and large, its areal strength is more in the foothills and tarai region, and the lower parts of the river valleys. Sugarcane occupying nearly 221 hectare is next important crop of this season. It covers about 34 per cent of the cropped land in kharif season or nearly 20 per cent of the gross cropped area. But its cultivation is confined to only 6 villages viz. Malsi, Shahpur, Rasoolpur, Dumka Bangar, Fatehpur and Motipur where it covers 93.0, 77.76, 78.68, 14.70, 12.68 and 0.40 ha respectively. The cultivation of sugarcane is mainly confined in Hardwar and US Nagar districts because of favourable physical and techno-economic conditions. In the remaining 4 districts its areal strength is zero. Madua is third major crop of the state. It is widely grown all over the state and occupies about 38 hectares or nearly 5 per cent of the total cropped land in kharif season. Its areal coverage is not much in the agricultural land of the households. There are only 3 villages viz. Syunsi, Sheela Mala and Deonagar, where it covers more than 3 hectares of agricultural land of the households. Rajma occupying about 6 ha or 0.87 per cent of the total kharif cropped land is the other important crop of kharif season. It is grown by the households of only 3 villages viz. Barhgaon, Syunsi and Mukhwa. Jhingora, pulses, soyabean and other crops which occupy less than one percent of kharif cropped land. Besides these, there are other crops such as small millets, maize, pulses, vegetables which together occupy nearly 6.5 ha or 0.91 per cent of the kharif cropped land.

7.3.3.2 Cropping Pattern in Rabi Season

Rabi is the next important agricultural season which covers 36 per cent of the gross cropped area. Wheat, oilseeds and barley are the major crops of this season (Table 7.11). A brief account of each is given here.

Table 7.11: Village-wise Patterns of Landuse in Rabi Season

Crops	*Area (Hec.)*	*% to Total Area*
1. Wheat	391.15	96.48
2. Oilseeds	8.8	2.17
3. Barley	5.46	1.35
Total	**405.41**	**100.00**

Source: Compiled and Computed by the author.

Wheat is a major stepple crop here and is grown in almost all the villages. It is generally sown on the level fertile alluvial plain. It requires a lot of fertilizer and irrigation. It is grown over nearly 383.5 ha amounting to about 95 per cent of rabi cropped land or about 34 per cent of the gross cropped area. It is widely grown in all the villages except Mukhwa where the climatic condition is too harsh for its cultivation. Its hectorage strength ranges from as low as 2.18 ha in village Gethiya to as high as 116.85 in Malsa village. However, there are 17 villages where less than 5 ha land is devoted to the cultivation of wheat. Most of these villages lie in northern and central part of the state. In 6 villages wheat covers between 5 to 10 ha of rabi cropped land, while in 6 villages it occupies more than 10 hectares. Its maximum concentration is in Malsi (116.85 ha), Shahpur (74.80 ha), Rasoolpur (24.88 ha), Fatehpur (23.84), Devnagar (25.50 ha) and Motipur (20.80 ha). All these villages lie in the tarai region of the state where besides physical conditions like relief, climate and soil, techno-economic conditions are also very conducive for cultivation of wheat. The farmers are also very innovative and enterprising (Fig. 7.11).

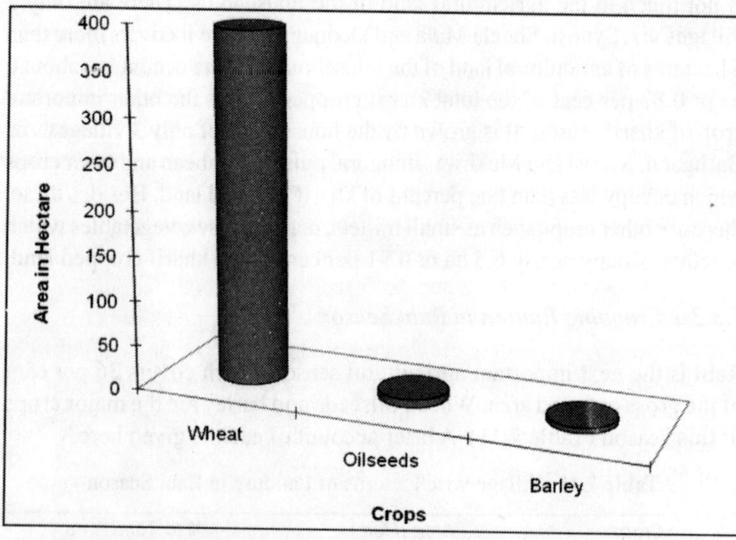

Fig. 7.11: Village-wise Pattern of Landuse in Rabi Season : 2005-06

Oilseeds is the next important rabi crop. It occupies nearly 9 ha of rabi cropped land accounting about 2 per cent of the total. Variety of

Photo 7.2: Farmers ploughing land in hilly areas

oilseeds are grown by households in comparatively less fertile land. It is quite often cultivated in combination with other crops like wheat, peas, gram etc. Oilseeds are grown only in 8 villages which are mostly located in the lesser Himalayan region.

Barley in the third important rabi crop. It is generally grown on less fertile soil found in the upper reaches of the villages. It also does not require much irrigation and fertilizers. It occupies about 6 hectares amounting to more than one per cent of the rabi cropped land. Like oilseeds it is also not very widely grown crop in the state. The households of only 9 villages have cultivated this crop. Its largest area of about 2 ha is found in Gethiya village. In the remaining villages it covers less than even one hectare. Its yield is also low and its contribution is very less in the status of food security in the villages.

Besides these, there are other crops like peas, potato, gram, pulses vegetables, etc. which are grown by the households but their hectorage strength is too meager to mention separately for each. They together occupy about 8 ha accounting to 2 per cent of the rabi cropped area. Because of their high nutritive value, their contribution in the status of food security is comparatively high.

7.4 Agricultural Diversification

The induction of new agricultural technology such as new farm implements, chemical fertilizers, improved varieties of seed, improved irrigation facilities, insecticides and pesticides, in the name of Green Revolution have no doubt increased agricultural production but it has created several problems such as land degradation, deforestation, water-logging, sinking of groundwater level, deteriorating soil fertility, contamination of groundwater and so on (Thakur, 2011). Therefore, the farmers are diverting their attention towards more sustainable agriculture. In this perspective diversification of agriculture is a better choice (Roy and Sharma, 2010; Pathak, 2002 and Singh, 2007). Diversification of agriculture includes two components i.e. diversification of cropping pattern and diversification of agricultural activities. The former includes shift from high soil exhaustive crops to less soil exhaustive indigenous crops, proper rotation of crop, cultivation of several crops on the same field over a period of time, while the latter includes the adoption of dairy farming, vegetable farming, horticulture, poultry farming, fish farming, apiculture, floriculture, piggeries, etc. The diversification of cropping

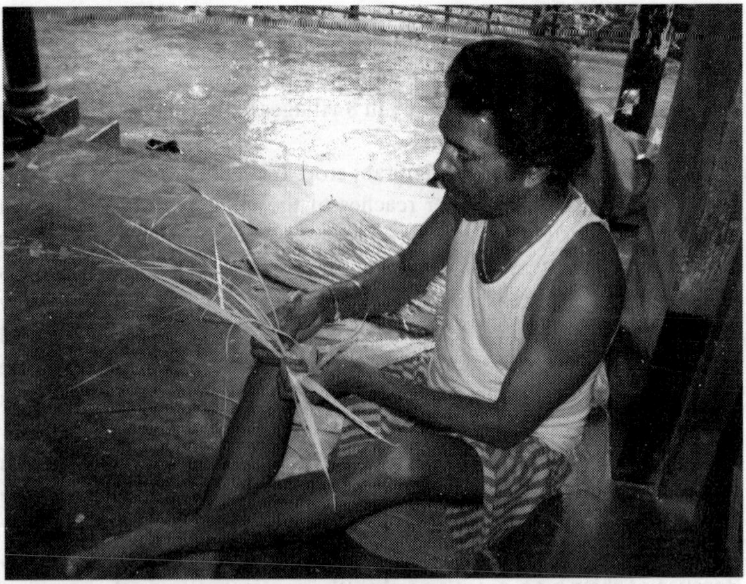

Photo 7.3: Cottage industry: An example of diversification

pattern has already been discussed along with the spatial patterns of agricultural landuse during kharif and rabi seasons but the diversification of agricultural activities has been briefly discussed and analysed here.

In view of the awakening of the farmers across the country regarding the diversification of agricultural activities, the selected households of the state have also diversified their agricultural activities and they have gone for many allied agricultural activities as is represented in Table 7.12.

Table 7.12: Agricultural Diversification: 2005-05

S.No.	Agricultural diversification	No. of respondents	% to total
1.	Dairy Farming	406	75.46
2.	Aquaculture	6	1.12
3.	Poultry Farming	7	1.3
4.	Apiculture	33	6.13
5.	Vegetable Farming	265	49.26
6.	Floriculture	4	0.74
7.	Horticulture	186	34.57
	Total Resp.	**907**	**100.00**

Source: Compiled and Computed by the author.

Table 7.12 shows that the households have diversified their agricultural activities in as many as 7 areas viz. dairy farming, vegetable farming, horticulture, poultry farming, aquaculture, apicullure and floriculture. These are respondents who have adopted more than one activity. Hence, the number of total exceeds much more than their actual number. Among these dairy farming is most popular and has been practiced by 406 households amounting to more than three-fourths of the total households. It is practiced on various bases and purposes. Vegetable farming which is required by almost everybody in one form or the other, is next very common type of farming which is practiced by 265 households accounting nearly half of the households. Both these activities are practiced across the state with few exceptions. Horticulture is the third important activity which is gaining popularity day by day. Its future is very bright in this hilly state. There are 186 households covering more than one-third of the total households who are practicing horticulture. Horticulture does not include only plantation of fruit plants but it also covers the medicinal and aromatic plants. Apiculture i.e. rearing of bees for honey and wax purposes is another important activity which is gaining popularity due to its remunerative medicinal values. It is also important because of its nutritional and medicinal value. There are 33

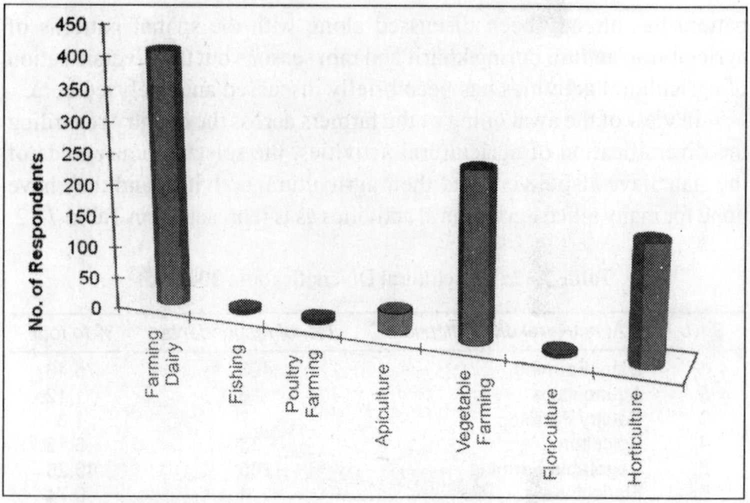

Fig. 7.12: Agricultural Diversification: 2005-06

households amounting to 6 per cent of the total households who are practicing apiculture. Poultry farming and acquaculture is not very common practices, perhaps, due to religious taboos. So is the case with floriculture which should have flourished much more than what it is because of its growing demand in the plain and abroad and its remunerative prices (Appendix-XIX).

The above analysis represents an aggregate picture of diversification of agricultural activities but there are a lot inter and intra-village variations in it. Therefore, a brief account of each activity is represented here.

7.4.1 Dairy Farming

Dairy farming is most important component of Indian agriculture. It has been encouraged during the British rule by establishing different dairy farms in order to meet the demand of dairy products for army. After independence India has also given much emphasis for dairy development during First Five Year Plan but real development has taken place after 'Operation Flood' on 'Anand Model' of dairy development. The White Revolution has taken place in three time period viz. 1970-80, 1981-85 and 1986 onwards- and it has continuously developed as a result of different programmes and policies of the government enacted during these periods (Shrivastava and Singh, 2008 and Singh and Akhtar, 2007).

Photo 7.4: Project Fallow Administering Questionnaire

Uttarakhand has huge potential for dairy development which provides employment to 1.5 lakh people directly or indirectly. Here dairy farming is developed on Anand Model at three levels, i.e., village—Milk Production Co-operative Societies, districts—Milk Production Co-operative Association and State—Uttarakhand Co-operative Dairy Federation. In the state various projects have been started for its development such as Intensive Mini Dairy Project, Fodder Bank Project, Anchal Milk Foods Ltd. etc.

In spite of all these efforts dairy farming has not developed uniformly across the state due to varied physical, techno-economic and socio-cultural environment (Appendix-XX). There are 4 villages where less than 50 per cent households are practicing dairy farming, while in 9 villages 50 to 80 per cent households have adopted this occupation. In the remaining 17 villages more than 80 per cent respondent are practicing dairy farming. Dairy farming, by and large, is practiced more in lesser Himalaya, Siwaliks and foothills.

Dairy farming is practiced in the state for different purposes and bases as is shown in Table 7.13 which shows that out of 406 dairy farmers, only 133 farmers amounting to nearly one-third of total dairy farmers practice it on commercial basis, while 273 dairy farmers accounting about 67 per cent of dairy farmers practice it on subsistence basis.

Photo 7.5: Uttarkashi: Nomads Moving Alongwith Their Animals

Photo 7.6: Preservation of Dry Fodder on Tree

Photo 7.7: Ladies Bringing Green Fodder for Animals

Table 7.13: Characteristics of Dairy Farming

No. of Households-406

Purpose	*No. of respondents*	*%age of total*
Commercial	133	32.76
Subsidiary	273	67.24
Main source	22	5.42
Subsidiary source	111	27.34
Personal use	273	67.24

Source: Compiled and Computed by the author.

But here also there are 8 villages where none of the households have adopted it on commercial basis, while in 13 villages less than 40 per cent households are doing it on commercial basis. In 4 vilages 40 to 60 per cent households and in 5 villages more than 60 per cent households are practicing dairy farming on commercial basis. Majority of households are practicing dairy farming on subsistence basis. In 8 villages 100 per cent households, while in 13 villages more than 60 per cent households

Photo 7.8: Dairy Farming

are doing it on subsistence basis. In 4 villages 40-60 per cent households and in 5 villages less than 40 per cent households are practicing dairy farming on subsistence basis. Substantial number of dairy farmers amounting to more than two-thirds of the total are doing it for personal use, while it is a subsidiary source of income for 111 dairy farmers accounting nearly 27 per cent of the total. It is a main source of income for remaining 22 dairy farmers who constitute about 5 per cent of the total.

Dairy farming is substantially practiced as a subsidiary source of income. In 9 villages, 100 per cent households practice it as subsidiary source of income, while in 8 villages more than 80 per cent households are doing it on the same basis. In 4 villages 60-80 per cent households practice it as subsidiary source, while in 9 village it is practiced for personal use. Dairy farming, by and large, is not practiced as main source of income. In 19 villages none of the households practiced it as main source, while in 10 villages less than 40 per cent households are doing it on this basis. Only in one village (Thewa) 100 per cent households are doing it as main source of income (Appendix-XX).

Photo 7.9: Champawat: Dairy Farmers Selling Milk

7.4.2 Vegetable Farming

India is the largest producer of vegetables in the world after China. In 2005-06 it had produced about 90 million tons of vegetables over an area of about 60 lakh hectares which is nearly 13 per cent of the worlds vegetable production. Variety of vegetables such as potato, brinjal, tomato, onion, lady's finger etc. are grown due to favourable physical, economic and socio-cultural factors. They are grown on small farms, garden, kitchen, gardens, roof tops, etc. (Chattoo *et al.,* 2009; Singh *et al.,* 2007). Uttarakhand also produces variety of vegetables. In 2007-08 it had produced 5,52,625 metric tonnes of vegetables over 56,854 ha of land. By and large, it is grown in Greater Himalayan and Middle Himalayan region.

There are altogether 308 households amounting to 57.20 per cent of the total households who are growing vegetables on various bases and purposes (Table 7.14). But there are a lot of inter and intra-village variations in it. There are 132 households amounting to nearly 43 per cent of the households who grow vegetables on commercial basis, while the remaining 176 households or 57 per cent of the total are doing it on subsistence basis. There are 12 villages where less than 40 per cent households are producing vegetables, while in 9 village 40 to 60 per cent households have gone for its cultivation. In 9 villages more than 60 per cent households are cultivating vegetables.

Table 7.14: Characteristics of Vegetable Farming

No. of Households- 308

Purpose	*No. of respondents*	*%age of total*
Commercial	132	42.86
Subsistence	176	57.14
Main source	44	14.28
Subsidiary source	111	30.04
Personal use	143	49.68

Source: Compiled and Computed by the author

Out of these, in 3 villages viz. Bangar, Sheela Mala and Syunsi 100 per cent households are producing vegetable on one basis or the other. In 14 villages less than 40 per cent households are practicing it on commercial basis, while in 9 villages the same percentage of households are cultivating vegetables on subsistence basis. In 7 villages 40 to 60 per cent households are growing vegetables on subsistence basis and in 14 villages more than 60 per cent households are producing vegetables on the same basis (Appendix-XXI).

Vegetable farming is not the main source of income for majority of the vegetable growers. In 13 villages it is not a main source of income even for one households, while in 11 villages less than 40 per cent and in 6 villages more than 40 per cent households are producing vegetables as their main source of income. Substantial number of households are doing it as their subsidiary source. In 13 village all the vegetable growing households are producing it as their subsidiary source of income, while in 5 villages less than 60 per cent households are growing it as subsidiary source. In remaining 12 villages 60 to 90 per cent households practice it as subsidiary source. From the above discussion, it may be inferred that vegetable farming is more popular in Lesser Himalaya and Siwaliks (Appendix-XXI).

7.4.3 Horticulture

Horticulture is an important component of Indian agriculture. It is grown on 4.1 crore ha of agricultural land which constitute about 13 per cent of it. Apple, mango, banana, citrus fruits, grapes, etc. are important fruits. Besides these, medicinal and aromatic plants are also grown. Uttarakhand is one of the major fruit producing state. Its production on commercial basis had been encouraged by Britishers in 1869 but it has got more impetus after insurgency in Kashmir valley. The state government has encouraged horticulture by instituting Directorate of Horticulture and by opening various horticulture research institutes. In 2007-08 the state had produced 7,39,015 metric tonnes of fruits over 19,04,60 ha of land which is nearly one-fourth of the net sown area. Many studies have been conducted to cover several aspects of horticulture but specific studies on status and future prospects of horticulture in Uttarakhand are lacking (Misger, Kumar and Masood, 2009; Sethi, Sharma and Masood, 2009). However, certain studies pertaining to medicinal plants and aromatic plants cultivation have been conducted by some scholars (Hamilton, 2003; Pande, 2007; Siddiqui, 2009 and Kaul and Dhar, 2007).

Horticulture is widely practiced across the state but is more concentrated in Greater Himalaya and Lesser Himalaya. However, there are 211 households amounting to nearly 39 per cent of the total households who grow fruits, medicinal and aromatic plans but their distributional pattern is not uniform in all the selected villages (Appendix XXII). There

Photo 7.10: A View of Horticulture plants in Mountainous Areas

are 20 villages where less than 40 per cent households are practicing it, while in 3 villages it is being practiced by 40 to 60 per cent households. In 7 villages more than 60 per cent households are growing horticultural crops.

Horticulture is a commercial type of farming and in majority of villages it is practiced on this basis. Out of 211 horticultural farmers 121 amounting to 57 per cent are doing it on commercial basis, while the remaining 90 households or 43 per cent are practicing it on subsistence basis. But it is not practiced uniformly in all the villages. There are 9 villages where more than 60 per cent households are doing it on commercial basis, while in 8 village about 40 to 60 per cent households practice it on commercial basis. In 10 villages less than 40 per cent households have adopted horticulture on commercial basis, while in 3 villages none of the households are doing it on commercial basis. A glance at the Appendix-XXII shows that there is no specific spatial pattern of practicing horticulture on commercial basis. However, by and large, in plains and lesser Himalaya large number of households are doing it for commercial purposes.

Photo 7.11: Horticulture: A View of Medicinal Plant

Horticulture is also practiced on subsidiary basis. There are 13 villages where more than 60 per cent households have adopted it on subsidiary basis, while in 6 and 7 villages 40 to 60 and less than 40 per cent households respectively are doing it on subsidiary basis. In 3 villages none of the households are doing it on this basis.

7.4.4 Floriculture

Flowers had special importance in socio-cultural and religious milieu in India since time immemorial but it has taken the place of an industry in the modern time in the light of its increasing demand in the country and abroad. Floriculture is practiced for several purposes such as landscaping, beautification, ornamental and so on (Neelofar and Masood, 2009; Raj, 2005; Chang, 1999). Consequently in 2006-07 India has produced 83,000 metric tons of bouquet flowers and 270 million tonnes of cut flowers. But however, it is still not very popular in Uttarakhand which has very favourable condition for floriculture. Although, almost every farmers is growing flower plants for beautification and religious purposes in their kitchen garden or courtyard but the area is too meager to be recorded during the course of the field work. There are only 4 households who are growing flowers on one basis or the other. Out of this, only one households is doing it on commercial basis, and the remaining 3 are planting it on subsistence basis. Similarly, it is a subsidiary source of income for only one households and the remaining 3 are, by and large, not selling the flowers.

7.4.5 Apiculture

Apiculture is a small scale occupation which can easily be practiced for the sustenance of life. It produces honey which is very nutritive and remunerative and is used even in modern medicines. In Uttarakhand it is an old occupation. It is cheap and less labour requiring occupation. It has a huge potential in the mountainous state of Uttarakhand but it has not yet gained momentum here. Altogether 33 households amounting to about 5 per cent of the total have gone for apiculture and they have been found only in 6 villages. Sheela Mala and Langri are the villages where 91 and 25 per cent households have gone for it respectively, while in Kalika, Barhgaon, Dungri Partiyal and Loharkhet 3 (18.75%), 2 (10.53%), 2 (12.50%) and 1 (6.67%) households are practicing apiculture respectively. But I am sure that with the modern techniques of apiculture and its increasing demand in medicine, it will flourish more in Uttarakhand because the state has very conducive environmental conditions for it.

7.4.6 Aquaculture

Due to large number of perennial and non-perennial rivers, canals, ponds, lakes, reservoirs and extensive large coastal area, India has huge potentiality of fish farming, and its production is continuously increasing over the years due to modern technology and the loosing grip of the socio-cultural taboos. India produced 7.5 lakh tonnes of fish in 1950-51 which has increased to 71 lakh tonnes in 2007-08. India is a exporter of fish which is also continuously increasing. In 1950-51 it had exported fish of about 2 crore rupees which has increased to 7,620 crore rupees in 2007-08.

Uttarakhand having a dense network of rivers and large number of ponds, lakes, and reservoirs also has huge potential for aquaculture. At present rivers are flowing over 2,686 sq.km., reservoirs have covered 20,075 hectares, natural lakes occupy 297 ha, while village ponds cover 628 ha area in Uttarakhand. The state has constituted the 'Fishing Policy' and has enacted fishing act in 2002 for encouraging fish farming. Uttarakhand has produced 2,572 metric tonnes of fish in 2005-06 which has increased to 3,091 metric tonnes of about 13 crore rupees in 2007-08.

Photo 7.12: Aquaculture in US Nagar

However, it is not a very popular occupation in the state because of religious taboos. It is practiced only in 2 villages i.e. Motipur and Mudiyani. In Motipur which is predominantly Bengali inhabited village, 5 households amounting to one-third of the total are engaged in aquaculture farming. In this village 80 per cent of fish farmers are practicing it on commercial purposes and its is a main source of income for nearly 75 per cent of fish farmers. In Mudiyani only one households does it on commercial basis and it is his main sources of income.

7.4.7 *Poultry Farming*

Due to the socio-cultural taboos poultry farming had not developed much in the past but in the recent years it has developed tremendously on different levels on scientific lines. As a result the production of eggs has increased from 1,832 million in 1950-51 to 53,532 million in 2007-08 resulting into the availability of 42 eggs per head per year. In Uttarakhand, poultry farming has not developed much in spite of several government incentives. It is practiced only in 3 selected villages. In Mudiyani 4 households (22.27%), while in Motipur and Malsi, 2 and 1 households

Photo 7.13: A View of Poultry Farm in Mountainous Region

respectively have adopted this occupation. But interestingly 6 out of 7 households have adopted it on commercial basis and it is a main source of income for 2 households of Motipur.

The above analysis of diversification of agricultural activities clearly indicates that it has considerably diversified in the state. The farmers no longer are confined to only cultivation of crops but they have ventured in different allied activities. The spatial pattern of diversification is not uniform except a few activities but however, regional pattern has emerged in many activities making distinction among high mountains, low mountains and foothills and plains.

7.5 Levels of Food Security

The levels of food security has already been discussed at district level in the previous chapter which represents its aggregate picture. Hence, it is studied here at village and household level in order to get in depth status at micro level. It would have been better if it is discussed at individual level based on dietary survey of each member of the household in order to get much clearer picture but it was not possible in this project given the time and financial constraints. The household food security has been discussed here first because the village level security is based on the aggregation of the household food security results.

7.5.1 Household Food Security

Household food security has been discussed and analysed in : (a) Calorific value, and (b) Monetary value.

7.5.1.1 Household Food Security in Calorific Value

The household food security has been worked out on the basis of net availability of food and the total number of consumption units in each household. The gross availability of food is calculated on the basis of total edible agricultural produce in the household plus all the purchased edible minus all the sold edibles in household from different sources. It may be mathematically expressed as :

Gross food available = Agricultural produce + Purchased edibles-sold edibles

It has been converted into calorific value as all the agricultural edibles cannot be added or averaged. The result thus obtained, has been multiplied by the coefficient of production i.e., 0.832 which gives the net food available.

The consumption unit of each household is calculated by multiplying the total members of the household with co-efficient of consumption unit i.e. 0.773. The net food available has been divided by the number of consumption units of the household and 365 in order to get the availability of calories per consumption unit per day. Based on this result and standard requirement of calories, i.e. 2430 calories per consumption unit per day, the households have been classified into 5 categories (Table 7.15).

Table 7.15: Household Level of Food Security in Calorific Value

Levels of food security	*No. of households*	*[illegible]ge of total*
Severely Insecure	91	[illegible]
Insecure	174	[illegible]
Secure	146	2[illegible]
Moderately secure	73	13.58
Highly secure	54	10.01
Total	**538**	**100.00**

Source: Compiled and Computed by author based on field data.

Table 7.15 shows that there are altogether 265 households amounting to about 50 per cent of the total who are food insecure. Out of this 91 households or 16.92 per cent of the total are in severely insecure category, while 174 households or 32.34 per cent are under insecure category. There are 273 households amounting to nearly half of the total are secure from food point of view which is a good sign of prosperity. Out of this 146 households or 27 per cent are in the secure category, while 73 households or 13.6 per cent are moderately food secure. The remaining 54 households amounting to one-tenth of the total are high food secure. The distributional pattern of the households is not uniform in all the 30 selected villages. By and large, the villages located in the foothill, tarai and plain areas or lower reaches of the major river valleys have larger number of households under food secure category, while the villages of other areas specially in high mountainous areas have large number of households under insecure food category (Fig.7.13).

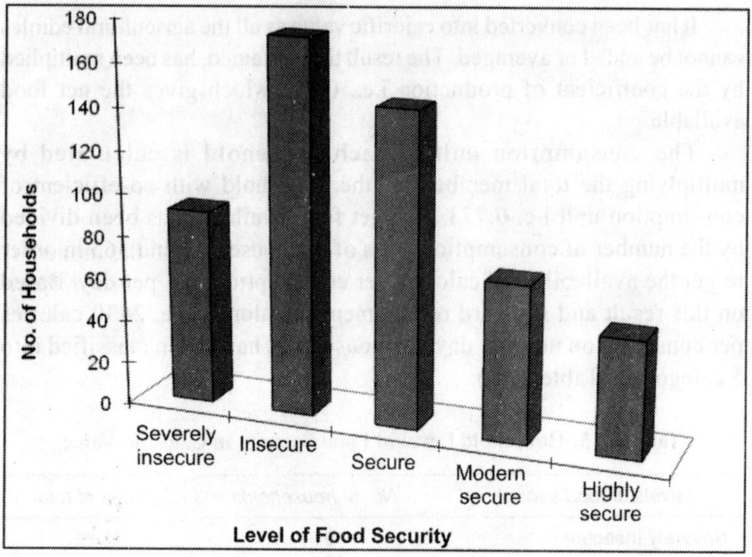

Fig. 7.13: Household Level of Food Security in Calorific Value

7.5.1.2 Household Food Security in Monetary Value

The household food security in value added terms has been worked out with the help of net food availability in value added terms and the consumptions units. Since, all the agricultural produce cannot be converted into calories, they have been converted into money value with the help of state level average prices of the different crops. In addition to this, the income of the household from other sources such as dairy farming, vegetable farming, horticulture, poultry, fishing, agriculture, floriculture, remittance, service, labour, etc. have also been added together. The result thus obtained has been multiplied with coefficient of production i.e. 0.832 which gives net food available. The net food available has been divided by the already calculated household consumption units in order to get the availability of money per consumption unit. Based on the results thus obtained the total number of households have been divided into 5 categories according to the standard requirement of fund to sustain one life for one year (Table 7.16).

Table 7.16 represents that altogether 253 households amounting to nearly 47 per cent of the total are food insecure. Out of this 97 households or 18.03 per cent are severely food insecure, while 156 households or

Table 7.16: Household Level of Food Security in Monetary Value

Levels of food security	*No. of households*	*%age of total*
Severely insecure	97	18.03
Insecure	156	29.00
Secure	103	19.14
Moderately secure	83	15.43
Highly secure	99	18.40
Total	**538**	**100.00**

Source: Compiled and Computed by the author based on field-data.

29 per cent are in insecure category. Most of these households are found in the high mountainous villages where besides agriculture, other sources of income are negligible. There are altogether 285 households amounting to nearly 53 per cent of the total are food secure. Out of this 103 households or 19 per cent are just food secure, while 83 households or about 15 per cent are moderately food secure. There are 99 households or about 18 per cent of the total who are highly food secure. These households are, by and large, found in the foothill, tarai, plain, and lower parts of the

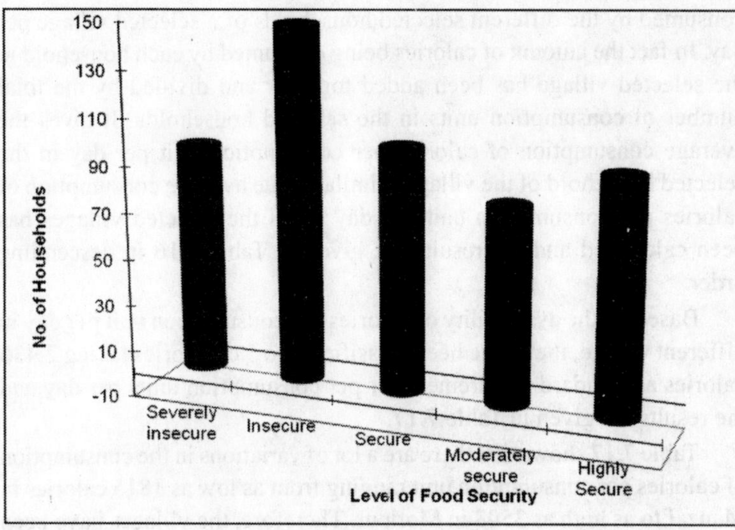

Fig. 7.14: Household Level of Food Security in Monetary Value

major river valleys. The households of the villages of the Greater Himalayas, and Middle Himalayas are, by and large, food insecure because of lack of other sources of income.

The comparative study of status of food security in calorific value and monetary value reveals that larger number of households (253) or (47%) are food insecure in value added terms mainly because of the lack of other sources of livelihood, while lesser number of households (265 or 49%) are food insecure in terms of calorific value. It shows that still the farmers are dependent on agricultural food production.

7.5.2 Village Level Food Security

The village level food security has been worked out for each selected village with the help of food security of the selected households. It has also been studied in two terms i.e. (a) Calorific Value; and (b) Monetary Value.

7.5.1.1 Village Level Food Security in Calorific Value

The village level food security is the average of the calories being consumed by the different selected households of a selected village per day. In fact the amount of calories being consumed by each household in the selected village has been added together and divided by the total number of consumption units in the selected households. It gives the average consumption of calories per consumption unit per day in the selected household of the village. Similarly, the average consumption of calories per consumption unit per day in all the selected villages has been calculated and the results are given in Table 7.16 in descending order.

Based on the availability of calories per consumption unit per day in different village, they have been classified into 5 categories taking 2,430 calories as standard requirement for per consumption units per day and the results are given in Table 7.17.

Table 7.17 shows that there are a lot of variations in the consumption of calories per consumption unit ranging from as low as 1815 calories in Manzaf to as high as 3508 in Motipur. Therefore, the villages have been grouped into 5 categories keeping the standard requirement of calories in view (Table 7.18)

Table 7.17: Village-wise Consumption of Calories

Villages	*No. of calories*	*Villages*	*No. of calories*
Manzaf	1815	Mukhwa	2689
Lohar Khet	1885	Langri	2695
Hitanu	1937	Bhatwari	2700
Arey	1965	Bangar	2738
Kalika	2110	Maletha	2780
Bargaon	2033	Syunsi	2787
Gethia	2180	Langasu	2815
Mudiyani	2257	Ratuda	2895
Guletti	2281	Fatehpur	2925
Kaffar Khan	2327	Shahpur	2957
Majkhali	2385	Rasoolpur	3015
Thewa	2452	Dumka Bangar	3023
Amsaur	2590	Devnagar	3281
Dunger Partiyal	2630	Malsi	3398
Sheela Mala	2675	Motipur	3508

Source: Compiled and Computed by the author.

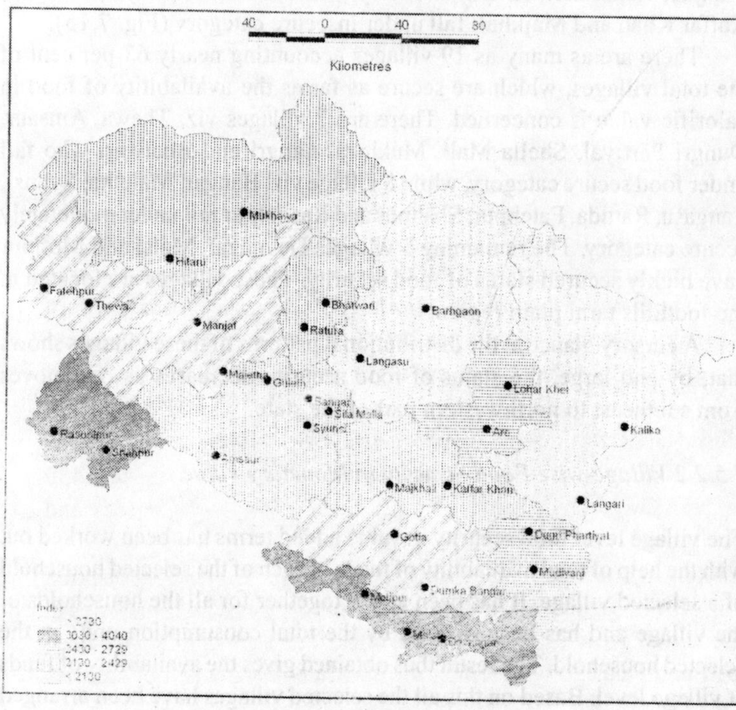

Fig. 7.15: Level of Food Security in Calorific Value

Table 7.18: Spatial Pattern of Level of Food Security in Calorific Value

Food security level	*Calorific range*	*No. of villages*	*% to total*
Severely insecure	< 2130	6	20.00
Insecure	2130-2429	5	16.70
Secure	2430-2729	7	23.30
Moderately secure	2730-3030	9	30.00
Highly secure	> 3030	3	10.00
Total		**30**	**100.00**

Source: Compiled and Computed by the author.

Table 7.18 shows that there 11 villages amounting to about 37 per cent of total which are insecure with respect to the availability of calories per head per day. Out of this 6 villages are in severely insecure category, while 5 villages amounting to 16.7 per cent of the total villages fall under insecure category. The severely insecure category villages are Kalika, Manjab, Loharkhet, Hitanu and Arey, while Gethia, Mudiyani, Guleth, Kaffar Khan and Majkhali fall under insecure category (Fig. 7.15).

There are as many as 19 villages accounting nearly 63 per cent of the total villages, which are secure as far as the availability of food in calorific value is concerned. There are 7 villages viz. Thewa, Amsaur, Dungri Partiyal, Shella Mali, Mukhwa, Langri and Bhatwari who fall under food secure category, while 8 village viz. Bangar, Maletha, Syunsi, Langasu, Ratuda, Fatehpur, Shahpur and Rasoolpur fall under moderately secure category. The remaining 3 villages Devnagar, Malsi and Motipur have highly secured status of food security. These villages are located in the foothills tarai plain region.

A cursory glance at the distributional patterns of these villages shows that, by and large, the status of food security decreases as one moves from southeast to northwestern part of the state.

7.5.2.2 Village-wise Food Security in Monetary Value

The village level food security in value added terms has been worked out with the help of total availability of funds in each of the selected household of a selected village. It has been added together for all the households of the village and has been divided by the total consumption units in the selected household. The result thus obtained gives the availability of funds at village level. Based on this all the selected villages have been arranged in ascending order (Table 7.19).

Table 7.19: Village-wise Availability of Monetary Funds

Villages	*Monetary Value (Rs.)*	*Villages*	*Monetary Value (Rs.)*
Manzaf	1,815	Syunsi	13,582
Arey	7,998	Sheela Malla	13,711
Hitanu	8,248	Bangar	13,825
Loharkhet	8,570	Bhatwari	13,852
Barhgaon	8,907	Langar	13,887
Kalika	9,667	Mukhwa	13,987
Gethia	9,847	Langasu	13,987
Mudiyani	10,125	Ratuda	14,026
Kaffar Khan	10,237	Fatehpur	14,512
Guleth	10,272	Shahpur	14,785
Majkhali	10,310	Dumka Bangar	16,101
Thewa	11,900	Rasoolpur	17,210
Amsaur	12,387	Devnagar	18,527
Dungry Partiyal	12,539	Motipur	20,578
Langary	12,813	Malsi	21,265

Source : Compiled and Computed by the author.

Table 7.19 indicates considerable variations in the availability of funds in the households of the selected villages. It ranges from as low as Rs. 1,815 in Manzaf Village to as high as Rs.21,265 in Malsi village. Therefore, the villages have been classified into 5 categories based on availability of funds and its standard requirement (Table 7.20).

Table 7.20 shows that there are 11 village which are food insecure with less than Rs.11000 per consumption unit per annum. Out of this 5 villages fall in the severely food insecure region, while in 6 villages, the households are food insecure. These food insecure village are widely distributed across the state but, by and large, their numbers is comparatively larger in Greater Himalayas and northern part of the Middle Himalayas.

Table 7.20 further shows that 19 villages are food secure. The main reason for it are favourable condition for agriculture, commercial cash cropping, and remittance from within and outside the state. Out of this 4 villages viz. Thewa, Amsaur, Dungri Partyal and Langari fall in the food secure region, while 5 villages viz. Dhumka Bangar, Rasoolpur, Devnagar, Motipur and Malsi fall under highly food secure region. These are the villages located in the foothills, tarai, and plains. Here, basmati rice, wheat and sugarcane are the main cash crops which increase the availability of funds in the villages. The remaining 10 villages fall under moderately food secure region. They are widely distributed in the state but their higher concentration is in the Middle Himalayan region.

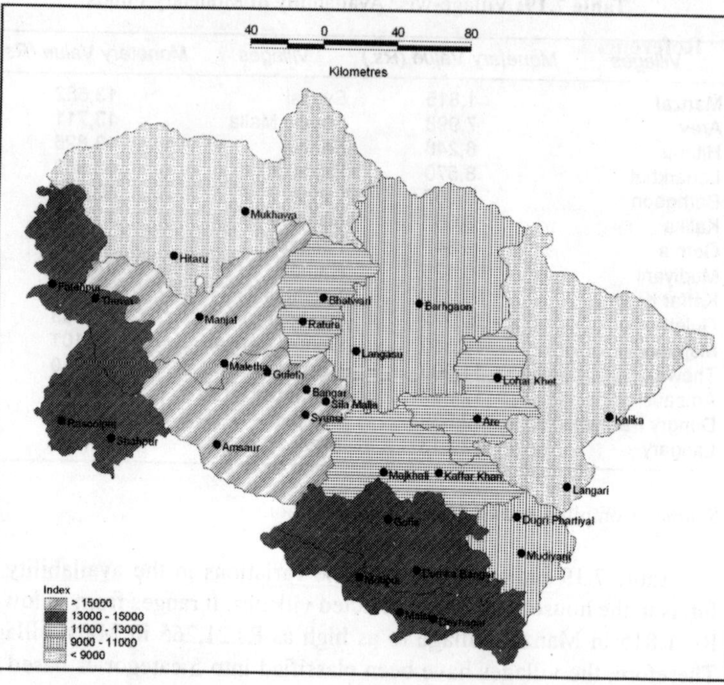

Fig. 7.16: Level of Food Security in Monetary Value

Table 7.20: Spatial Pattern of Level of Food Security in Monetary Value

Food Security Level	*Monetary Range*	*No. of Villages*	*% to Total*
Severely insecure	< 9000	5	16.70
Insecure	9000-10999	6	20.00
Secure	11000-12999	4	13.30
Moderately secure	13000-15000	10	33.30
Highly Secure	> 15000	5	16.70
Total		**30**	**100.00**

Source: Compiled and Computed by the author.

A comparative analysis of food security in calorific value and in value added terms indicates that by and large, the households an more secure in value added terms in comparison to calorific value mainly due to income from other sources rather than agriculture alone.

References

Bhatt, J.P., 2003 : 'Prospective Fisheries Development in Uttaranchal State', in Rawat, M.S.S. (Ed.), *Central Himalaya : Environment and Development*, Department. of Geography, H.N.B. Garhwal University, Srinagar.

Bisht, D.S., 2006 : *Poverty, Planning and Developing : A Case Study of Uttarakhand State*, Central Himalayan Institute, Dehradun.

Bose, S.P. 1961 : Characteristics of Farmers who Adopt Agricultural Practices in Indian Villages, *Rural Sociology*, Vol.26.

Chang, S.T., 1999 : Global Impact of Edible and Medicinal Mushroom on Human Welfare in 21st Century, *International Journal of Medicinal Mushroom*, Vol. I.

Chatto, M.A. *et al.*, 2009 : 'Vegetable Gardening' in Masood, S. (Ed.), *Peri-Urban Agriculture : Agriculture by Twenty 20 and Beyond*, A.P.H. Pub. Corp., New Delhi.

Chikara, D.V., 2006 : "Levels of Agricultural Development in Haryana" (Unpublished Ph.D. Thesis, Geography Deptt., Delhi University, Delhi.

Hamilton, A., 2003 : *Medicinal Plants and Conservation : Issues and Approaches*, International Plant Conservation Unit, U.K.

Kaul, T.N. and Dhar, B.L. 2007 : *Biology of Cultivation of Edible Mushroom*, Westville Pub. House, New Delhi.

Lee, E.S., 1966 : A Theory of Migration, *Demography*, Vol.3.

Misger, F.A., Kumar, A., and Masood, S., 2009 : 'Water and Fruit Crops' in Masood, S. (Ed.) *Peri-Urban Agricultural : Agriculture by Twenty 20 and Beyond*, A.H.P. Pub. Corporation, New Delhi.

Mohammad, N.. 1978. Impact of Social Factors on Diffusion of Agricultural Innovations in Central Trans-Ghaghara Plan, *National Geographical Journal of India*, Vol.XXIV, Varanasi.

Mohammad, N.. 1981 : *Perspectives in Agricultural Geography*, Vol.I to V, Concept Pub. Co., New Delhi.

Mohammad, N.. 1988 : *Caste and Primary Occupation: A Geographical Analysis*, Concept Pub. Co., New Delhi.

Mohammad, N.. and Majeed, A., 1979. 'Socio-economic Factors and Diffusion of Agricultural Innovations' in Mohammad, A. (Ed.), *Dynamics of Agricultural Development in India*, Concept Pub. Co., New Delhi.

Neelofar, L. and Masood, S., 2009 : 'Floriculture, Landscaping and Recreation' in Masood, S. (ed.), *Peri-Urban Agriculture : Agriculture by Twenty 20 and Beyond*, A.P.H. Pub. Corp., New Delhi.

Pande, P.C., 2007 : The Value of Traditional Medicinal and Aromatic Plants of Uttaranchal, *Proceeding of V.C.'s Conference*, Nainital.

Pathak, R.K., 2002 : 'Diversification under Different Agro-ecosystem' in Sharma, A.K. (Ed.), *Agricultural Diversification : Problems and Prospects*, I.K. International Pub. House, Pvt. Ltd., New Delhi.

Radha, V.. 2009 : *Women and Rural Household Food Security*, Serials Publications, New Delhi.

Raj, D., 2005 : 'Bio-Aesthetic Planning' in Bhattacharjee, S.K. (Ed.), *Advances in Floriculture*, Pointer Pub., Jaipur.

Ramachandran, R., 1969 : *Diffusion of Agricultural Innovations in Rural India*, Mysore.

Ramachandran, R., 1992 : 'Correlates of Innovativeness among Farmers in the Ganga Plain' in Mohammad, N (Ed.) *Human Perception and Technological Change in Agriculture,* Concept Pub. Co., New Delhi.

Rangaswamy, P. *et al.,* 1972 : *India's Changing Farmers*, Allahabad.

Roy, P.K. and Sharma (Eds.), 2010 : *Globalisation and Diversification in India*, Regal Pub., New Delhi.

Sethi, S. Sharma, R.R. and Masood, S., 2009 : 'Value Addition of Fruits', in Masood, S. (Ed.), *Peri-Urban-Agriculture : Agriculture by Twenty 20 and Beyond*, A.P.H. Pub. Corp., New Delhi.

Siddiqui, M.A.A. and Masood, S., 2009 : Medicinal and Aromatic Plants' in Masood, S. (Ed.), *Peri-Urban Agriculture : Agriculture by Twenty 20 and Beyond*, A.P.H. Pub., Corp., New Delhi.

Singh, H. and Akhtar, R., 2007 : 'Strategies for Dairy Development in Kashmir Valley : A Geographical Perspectives', in Thakur, B. (Ed.), *City, Society and Planning*, Concept Pub. Co., New Delhi.

Singh, N.P. *et al.,* 2004 : *Modern Technology on Vegetable Production*, International Book Distributing Co., Lucknow.

Singh, R.P., 2007 : 'Role of Bio-fertilizer in Crop Production' in Sharma, A. Singh, R.K. (Eds.), *Composite Farming Practice and Economic Development*, Abhijeet Pub., New Delhi.

Sinha, R.R.K., 2009 : *Dynamics of Land-Caste Relations in India*, Manak Pub. New Delhi.

Srivastava, M.P. and Singh, J.S., 2008 : *Dairy Development in the New Millennium*, Deep and Deep Pub. New Delhi.

Thakur, B., 2011 : *Depletion and Degradation of Land Resource of India*, R.N. Dubey Foundation, Allahabad.

Verma, K. 2007 : Role of Urban Agriculture in Life Support System, A Case Study of Alipur Block, Delhi (Unpublished M.Phil. Dissertation, Deptt. of Geography, Delhi University, Delhi).

Chapter 8

Summary Conclusion and Suggestions

The state of Uttarakhand which came into existence on 9th November, 2000 in accordance with Uttar Pradesh Reorganisation of State Act 2000, has been carved out from U.P. by taking its hilly region including Garhwal and Kumaon divisions. Subsequently, the district of Hardwar is also added to it. Sprawling over on area of 53,484 sq. km, it is situated between 28° 45′ 1″ to 31° 27′ 42″ N. lat. and 77° 34′ 10″ to 81° 92′ 11″ E long. It is a part of the youngest and highest mountain system of the world and is the true representative of the total Himalayan geographical system. It is a panorama of an intricate mosaic of mountain ranges, hills, valleys and plains. However, the state has two broad surface features i.e. the piedmont plains of Bhanbar and Tarai together with a small patch of Ganga plain in the south and the magnificent Himalayas in the north. These regions vary considerably in their geographical, genomic, bio-physical, socio-cultural and other geo-human characteristics. There are marked differences in their altitude, complex litho-tectonic features, distinct orographic alignment, slope, structure and composition of soil, climatic condition, natural vegetation and wide range of biodiversity. These variations have performed unique functions in shaping its varied natural and cultural landscape. It has a dominating role in limiting the habitation, wide range of means of livelihood, dimensions of socio-cultural and religious facts and unique arrangements of the state as a whole.

However, these high mountain ranges interspersed with ridges and river valleys provide little scope for agricultural activities. Only 13 per cent of the total area of the state is under cultivation which is substantially found in southern part of the state. Therefore, the agricultural area should be extended specially in river valleys without much disturbing the ecological balance. The heterogeneous topography provide conducive condition for allied agricultural activities specially dairy farming horticulture, forestry, etc. which should be encouraged.

The physical diversities and complexities have resulted into complex and contrasting climate within a short distance. There is a bewildering confusing varieties of climate in a single mountain range. The whole range of world climates from tropical to alpine are found if one moves from foothill plains to perpetually snow bound zone of Greater Himalayas. However, on the whole, the state experiences monsoon type of climate which is characterized by reversal of winds, seasonality, periodicity, variability and intensity of rainfall which adversely affects the agricultural system and its diversification. It is more so in the tarai and plain areas which are predominantly an agricultural tract. Therefore, various attempts should be made to make alternative arrangements such as development of minor irrigation projects which will not only provide irrigation but will also help in generating hydel power that could be used for domestic purposes, cottage to household industries, proper crop management practices and diversification of agriculture with reference to cropping system and agricultural activities.

The state has huge potential water resources in the form of running water and glaciers. It has a dense network of substantially perennial rivers and innumerable glaciers. Nearly 17 per cent area of Himalaya lies under perpetual snow. The Ganga, Yamuna, Kali, Sarju, Ramganga are the major drainage systems which provide huge volume of water that could be used for surface irrigation through gravity forces and generation of hydroelectricity. The state has conducive geographical condition for hydel-power projects. The glaciers are the major sources of water for many rivers of the state which not only provide water to the state during non-rainy season but also down the plain where it is a major source of irrigation. Therefore, this precious natural resource should be properly and scientifically utilized for the overall sustainable development of the state keeping the maintenance of ecological balance on top priority.

The nature has bestowed the state with precious forest resources which cover about 61 per cent area of the state. The areal coverage of forest varies from one-third in US Nagar to about 89 per cent in Uttarkashi district. It is the abode of innumerable varieties of flora and fauna. There is variety of forests ranging from tropical moist deciduous to dry alpine shrubs. They are indispensable sources of food, fodder, fibre, fuel, and flowers. But unfortunately, they are being cleared mainly for agricultural purposes specially in the tarai belt. It has created and continuously creating many environmental problems specially reducing water table, soil erosion, land degradation, shortage of drinking water and desertification. Therefore, a proper and effective forest policy should be evolved in the form of social forestry, agro-forestry, etc. for the proper management and conservation of forest resources. The Joint Forest Management Policy should be more effectively implemented. The peoples awakening and their participation in the conservation and management should be more encouraged with better and more profitable incentives.

The state has poor soil with the exception of fertile alluvial soil of tarai and plain. In remaining areas it is in the thin layer full of coarse sand, pebbles and stones. It is very fragile and is very susceptible to erosion and degradation. Therefore, the soil should be utilized with utmost care taking the physical, chemical and biological characteristics of soil into consideration. Proper arrangement should be made for the scientific testing of the soil of the farms of the farmers who should be given soil card of their fields indicating the properties of the soil. Based on these characteristics of the soils, scientific and rational landuse planning should be done for the conservation and management of soil. Proper rotation system of crops, a better cropping pattern and crop diversification should be adopted.

The state is not very densely populated except the tarai and plain areas. The average density of population in 89 persons per km^2. Keeping the potentiality of the resources and over all carrying capacity of land, the pressure of population is not very high. Therefore, the potential resources, should be developed to improve the overall condition of the people specially with reference to food security. The state is inhabited by a large number of tribes who are socially, economically and culturally backward. They, by and large, come as seasonal labourers in agricultural activities in the plain. Therefore, the specific development policy should be evolved for these tribal people. There are certain tribes who are still practicing nomadism. They should also be encouraged for sedentary life.

The general landuse pattern indicates that forest occupy substantial area amounting to nearly 61 per cent of the total area. Nearly 8 per cent area is not available for cultivation which includes 5.5 per cent under barren and uncultivated land. This land has the potentiality to be brought under various kinds of afforestation. The area also may be utilized for growing medicinal and aromatic plants. It may also be given to social forestry. Nearly 15 per cent of the state area is under uncultivated land excluding fallow land. Out of this, nearly 7 per cent is culturable waste which is lying unutilized for agricultural purposes due to one reason or the other. It has the maximum potentiality to be brought under cultivation or other more profitable uses. It should be reclaimed using different reclamation techniques. It will increase the agricultural land which will enhance the agricultural production leading to more food security. This land should not only be brought under crop cultivation but many also be given to forestry, fodder cultivation, pasture and horticulture which will ultimately help in maintaining food security in the state. Because of overall adverse physical environment specially relief, surface configuration, the amount of net sown area is only 5,66,567 ha amounting to only 13 per cent of the total area. This is not uniformly distributed and is not very fertile except in tarai and plain areas together with lower river valleys. It is a very fragile soil and is very susceptible to deteoriation and degradation. Therefore, it should be properly managed and conserved. The scientific conservation measures such as levelling and bunding, terracing along the contour, proper rotation of crops, ploughing and planting along the contour and cultivation of deep rooted crops should be adopted. The cropping system should be diversified.

In spite of huge potential water resources, only 45 per cent of the net area sown is irrigated and that too mainly by tube-wells (62%) and canals (28%). It shows that irrigation facilities have been developed more in the tarai and plain areas where boring of tube-well is easy and possible. The irrigation facilities in the state specially the canal irrigation through gravity forces should be developed and it should be scientifically and rationally utilized without creating the problem of land degradation. However, in the plain areas specially in Hardwar district, it has created the problem of land degradation.

The mountainous region of Uttarakhand with undulating relief and rugged topography leading to limited land available for cultivation, agriculture remains the basal activity of man. There are three main determinants of agriculture, i.e., availability of water, temperature condition and slope. The spatial distribution of cultivated land is basically the result of the complex interplay of these factors together with certain local factors right from the high altitude mountain environment close to snow ranges in the north down to the subtropical zone of tarai and plain in the extreme south.

Owing to the varied and complex ecological, techno-economic and socio-cultural factors, several systems of cultivation are practiced in the state. However, three special type of cultivation are practiced here which are locally known as talaon, upraon and katil. Talaon cultivation also known as low level or irrigated cultivation is generally practiced in the low lying areas of major mountain river valleys which are characterised by deep fertile alluvial soil with high moisture retention capacity, high humus content, high temperature condition, and abundance of irrigation water. These are the areas of intensive cultivation producing variety of crops like wheat, rice, vegetables, pulses and oilseeds. The upraon cultivation is practiced in a little higher areas of talaon cultivation. Here the soil is comparatively poor and irrigation facilities are meager. Hence, extensive cultivation with coarse grains like maize, madua, barley, jhingora, etc. are grown here. The yield is low. The Katil cultivation—also known as intermittent cultivation, is practiced in unirrigated upland areas with poor and infertile soil. Obviously, the yield is poor. These are ecologically more appropriate indigenous practices which should be encouraged more on scientific lines.

Wheat, rice and madua are the major crops occupying together nearly 65 per cent of the net sown area. They are followed by maize, vegetables and oilseeds each occupying nearly 3 per cent of net area sown. Besides, there are many other crops occupying less percentage, are grown in the state. The cropping pattern is not very rational and scientific. There are a lot of spatio-temporal variations in it. In certain districts, such as US Nagar, Hardwar and Dehradun, it is heading towards specialization and commercialization which will ultimately adversely affect the soil health leading to land degradation. Therefore, the cropping pattern should be diversified and more and more crops should be grown which will help in maintaining ecological balance, and will provide more nutritive and balanced diet.

Since land resource is meager and its demand is very high, it is utilized very intensively. The average intensity of cropping is 157 per cent which is much above the national average but there are districts like US Nagar and Nainital where intensity of cropping is more than 173 per cent. This high intensity is not very advisable as it will deteoriate land and make it unfit for cultivation in the long run. Therefore, a judicious and balance intensity of cropping should be followed so as to maintain the sustainability of land ecology.

The high intensity of cropping leads to cultivation of many crops which are grown in several associations. In the state a maximum of 5 crops come in combination which is not a better sign for the health of agricultural lands. Although 2 crop-combination has been found only in one district but 3 crop combination are found in 7 districts. Five crop-combinations have emerged in 4 districts, while only in one district 5 crop combination has been found. It shows that the cropping is not diversified at desired level. Therefore, programme and policies should be made to diversify the cropping pattern by cultivating more and more number of crops depending upon the geographical condition of the area.

As a result of this, crop diversification is low. There are only 2 districts where very high crop diversification index has emerged, while in 4 districts it has come out to be high diversification. In 4 districts low agricultural diversification has emerged, while in 3 districts it is moderate. This is not very encouraging result. The diverse cropping in the state, therefore, should be more encouraged and the larger number of crops should be grown which will not only help in maintaining soil status but it will also increase the level of food security.

Since the distributional pattern of population is very varied and ununiform which is the outcome of the interplay of varied and complex ecosystem, techno-economic and socio-cultural factors, the spatial distribution of consumption unit, obviously follow the same pattern. Lying in the fertile alluvial plain US Nagar and Hardwar have larger number of consumption units and are followed by Dehradun—the state capital district, and Nainital- the tourist and educational hub. These 2 districts put together account more than one-third of consumption units, while the latter 2 districts account nearly one-fourth of the consumption units. As such about 58 per cent consumption units are found in these four districts. The remaining 42 per cent are scattered in 9 remaining districts of the state. However, the general distributional pattern of the consumption units goes on decreasing as one moves from south west towards northeast. Therefore, it is suggested that more jobs and opportunities should be created in the north eastern part of the state so as to attract the population and controls their movement towards the plain.

The carrying capacity of land, which has been worked out with reference to calorific and monetary values using most appropriate technique of its measurement, indicates that Dehradun and Nainital have the highest carrying capacity in terms of standard nutrition unit because of their being administrative and educational centre respectively. It is followed by Hardwar and US Nagar which are basically agricultural districts falling in the fertile alluvial tract. In the remaining districts the carrying capacity is low mainly because of poor agricultural condition. Almost similar distributional pattern is noticed in case of carrying capacity of land with reference to monetary value but the number of SNU has gone up in almost all the districts because of other sources of income such as remittance, service and allied agricultural activities. Obviously in Nainital and Dehradun the other sources of income are much more as compared to other districts because these are administrative and tourist-cum- educational centres respectively. As such the people have more purchasing power. In the Hardwar and US Nagar it is high because of cultivation of more remunerative crops like sugarcane, export quality basmati rice, etc. In the remaining districts also, the situation of SNU is better because of remittances and cultivation of horticultural, medicinal and aromatic plants and other subsidiary sources of income. However, there is still much scope to increase the carrying capacity of land by adopting more rational technique of agriculture without much disturbing the ecological balance, developing eco-tourism and diversification of agriculture specially in case of diversification of agricultural activities. The status of food security has been ascertained in terms of calories and monetary values using most appropriate technique at district level using secondary data. In case of calorific value it has been found that it ranges from as low as 2113 calories in Chamoli district to as high as 3255 calories in US Nagar which is not a very encouraging figure. There are 6 districts (46%) which are food insecure. Out of this 2 districts are severally insecure and 4 are insecure. It has resulted in the occurrence of nutritional deficiency diseases as has been observed during the fieldwork. Nearly 54 per cent districts are food secure. Out of these 2 districts viz. US Nagar and Nainital are highly food secure. Unexpectedly Hardwar has not come in this category because of high density of population and more area devoted to sugarcane cultivation. Therefore, it is suggested that food grain based agriculture should be encouraged and the agriculture should be more diversified.

The status of food security in terms of monetary value is comparatively better because of inclusion of other sources of income and cultivation of remunerative cash crops. There are only 9 districts which are food insecure, while the remaining 4 districts are food secure. In the light of the above discussion it is suggested that the farmers should not depend only on cultivation of crops but they should diversify their agricultural activities in the form of dairy farming, vegetable cultivation, horticulture, apiculture, poultry farming, aquaculture, floriculture, etc. which will enhance their purchasing power for food and consequently increase the status of food security.

The core chapter—agricultural diversification and food security: a micro level analysis is entirely based on primary data. The agricultural diversification and food security assessed at micro level presents interesting results. The socio-economic structure of farmers and households reveal that the farmers are still, by and large, very traditional and orthodox. The elder farmers still dominate the decision making processes in agriculture and the younger generation are not taking keen interest in agriculture. They are looking for white collar jobs and better job opportunities within and outside the state. Although the general literacy level in the state is satisfactory but among the farmers it is low which in turn have adversely affected the process of agricultural development. The farmers still live in joint family system but interestingly the size of family is not very large. Perhaps, owing to low literacy rate, lesser exposure to outside world and unfavourable surface configuration, the level of innovativeness is low which in turn has adversely affected the growth of agriculture. Therefore, the farmers should be made more literate by implementing all the educational policies of the Central and State Government in letter and spirit. They should be made aware of various agricultural innovations through government and non-government agencies so that they adopt it which will enhance the process of agricultural development.

Patterns of landuse in Kharif season among the farmers does not represent very encouraging picture. They are highly concentrated on cultivation of foodgrains specially rice, wheat and madua. On the whole, the yield situation is poor leaving few exceptions such as in case of famers of US Nagar and Hardwar. They have not diversified their cropping pattern to the desired level which has not only adversely affected the health of the soil but also the nutritional status of the farmers. They are, therefore, advised to diversify their cropping pattern by including several crops which will ensure better crop yield in varied climatic conditions and also provide more calories to enhance their status of food security.

However, the encouraging result is that the farmers have diversified their agricultural activities. They have gone for as many as 7 areas for diversifying their agricultural activities. These are dairy farming, vegetable farming, horticulture, apiculture, floriculture, poultry farming and fisheries. Among these activities dairy farming is most popular. It is practiced by nearly 76 per cent of the farmers on various bases. Nearly one-third of the dairy farmers practice it on commercial basis which enhances their household income to a large extent, while remaining two- thirds practice it on subsistence basis. It is a main source of income for nearly 5 per cent of the farmers, while it is a subsidiary source of 28 per cent of the farmers. It is practiced substantially for personal use. In the light of the huge potentiality for dairy development in the state and its nutritive value, it is suggested that dairy farming should be developed in the state on priority basis specially by adopting Anand Model of Dairy Development. The pastures and grazing lands should be satisfactorily developed, managed and conserved. The farmers should be made conscious about its importance and economic potential through various extensive service programmes and also by NGOs who are working satisfactorily in the state.

Vegetable farming which is another area having huge potential in the state, is next important diversified activity engaging 50 per cent of the farmers. Vegetables are grown not only in agricultural fields but also in kitchen garden, courtyard, and in a small area in any corner of the agricultural field. It is developing fast due to refrigeration facility, and quick means of transport facility connecting the markets of the plain to the vegetable growing areas. It is also practiced on various bases. Nearly 43 per cent of vegetable growing farmers practice it for commercial purpose, while remaining 57 per cent are practicing it for subsistence purpose. It is the main source for about 14 per cent farmers, while it is subsidiary source for about 57 per cent of the vegetable growers. Nearly half of the vegetable growers are producing vegetables only for their personnel use. Whatever, may be the case, the growing of vegetable is gaining popularity in the light of its nutritive value, changing habit of the people towards being more vegetarian and increasing cost of the pulses. It is, therefore, suggested that the cultivation of vegetables should be further encouraged and it should be done on commercial basis and as a main source of income.

Horticulture is the third activity which is being practiced by the farmers. Nearly 35 per cent farmers are growing horticulture crops. It includes plantation of various types of fruits, medicinal and aromatic plants. The state has huge potential for it because of its varied climate and physiographic conditions. Majority of the farmers amounting to 60 per cent are doing it for commercial purposes, while remaining are practicing it as subsistence source of income. It is more nutritive and remunerative agricultural practice which will not only improve the health of the farmers but also will increase their purchasing power. Therefore, the horticulture farming should be encouraged and more and more area should be brought under it. Apiculture is another activity which has engaged about 6 per cent of the farmers. The state has the potentiality for it because of having variety of natural flowers, gardens and area under oil seeds which provide base for it. Floriculture is also gaining popularity in the state because of its great demand in the country and abroad. Besides these, poultry farming, and fishing are other activities being adopted by the farmers. The critical analysis of all these activities clearly indicates that the state has huge potentiality for it. Therefore, it should be encouraged by providing various kind of incentives to the farmers.

The study of level of food security has been analysed at household and village level with reference to calorific and monetary value. The study reveals that newly 50 per cent of the households are food insecure.

Out of this, one-third are severely food insure, while two-thirds are insecure. Out of food secure categories, nearly 27 per cent farmers are very high food secure, while only one-third are highly secure. The remaining 14 per cent are moderately secure. However, by and large, the status of food security is not satisfactory. Therefore, more and more food should be produced and purchasing power of the farmers should also be increased. The status of food security of household with reference to monetary value is better and satisfactory because the farmers have increased their purchasing power by adopting diversified activities. The distributional pattern of the farmers in indifferent categories of food secure is highly variable. About 47 per cent farmers are food insecure, while 53 per cent are food secure. It is suggested that the income of the farmers should be increased through diversification of their activities which will help in augmenting their purchasing power.

The village level food security has also been studied in calorific and monetary value. The study reveals that 11 villages amounting to 37 per cent of the villages are food insecure, while 63 per cent villages are food secure with reference to calorific value. The village as well as household food security is adversely affected by the Public Distribution System (PBS) which is defective and highly ineffective. The status of the villages in monetary value is not very different from the food security in calorific terms as far as the percentage of village in food secure and insecure categories are concerned but the distribution of villages with in the category is different. Large number of villages are in the moderate and high food security category. On the whole the status of food security in the state is not very satisfactory. Besides, the various factors discussed earlier, Public Distribution System is responsible to great extent. There is enough food in the stock of government but it is not properly distributed and it does not reach the poor people.

APPENDICES

APPENDIX-I

Questionnaire for Household Survey

Schedule No.:..........................
Date:......................................

Name of Village .... Block .... Tahsils ....District ....

1. Name of Village .... Age .... Caste .... Religion ....
2. Education: (a) Informal .... (b) Formal ....
3. Total members of household ....

Age Group: Years	*Less than 5*		*5 -15*		*16-30*		*31-60*		*>60*	
	M	F	M	F	M	F	M	F	M	F
Members:										
Educational Status										

4. Agricultural Landholdings: (Area in ....)

Type of Holding	*Irrigated*	*Unirrigated*	*Total*
Land Owned			
Land Leased in (i) Share (ii) Rental	....	....	....
Land Leased out (i) Share (ii) Rental	....	....	....

5. Pattern of Landuse in Kharif Season: (Area in ....)

Major Crops	*Irrigated/ Unirrigated*	*Source of Irrigation*	*Average Yield per unit*
....	....	....	....
....	....	....	....
....	....	....	....
....	....	....	....

6. Pattern of Landuse in Rabi Season: (Area in ....)

Major Crops	*Irrigated/Un-irrigated*	*Source of Irrigation*	*Average Yield per Unit*
....	....	....	....
....	....	....	....
....	....	....	....
....	....	....	....

7. Has your cropping pattern changed during last 5-10 years? Yes/No If yes, give details ....
8. Which crops you were growing before? ....
   (a) Do you practice mixed Farming? Yes/No, If Yes, What are the commodities of crops? ....
9. Has the yield of your crops – increased/remained the same/decreased?
10. Adoption of Agricultural Innovation

(a)

Agricultural Implements	*Owned/Hired*	*Areal Coverage*	*Operational Cost*

(b)

Chemical Fertilizers	*Kharif Season*				*Rabi Season*			
	Amount	*Crops*	*Area*	*Approximate Cost*	*Amount*	*Crops*	*Area*	*Approximate Cost*
....	.....	.....	.....	.....	.....	.....	...	.....
....	.....	.....	.....	.....	.....	.....	...	.....
....	.....	.....	.....	.....	.....	.....	...	.....

(c) *New Varieties of seed*	*Kharif Season (Area)*	*Rabi Season (Area)*
....	....	....
....	....	....
....	....	....

(d) *Insecticides and Pesticides*	*Kharif Season*			*Rabi Season*		
	Amount	*Crop*	*Area*	*Amount*	*Crop*	*Area*
....	....	....	....	....	....	....
....	....	....	....	....	....	....
....	....	....	....	....	....	....

Agricultural Diversification

I. *Dairy Farming*

(1) Do you practice dairy farming? Yes/No, If yes, since when? ……………………..

(2) *Animals*	*Number*	*Age*	*Amount of Milk*	*Milching Period*	*Rate of Sale*
Cows	………	………	………	………	………
Buffaloes	………	………	………	………	………
	………	………	………	………	………

(3) Commercial/Subsistence basis
(4) Main source/Subsidiary source
(5) Average Income per month………..
(6) Place of sale of milk and milk products ………………
(7) Do you also purchase and sell milk? Yes/No, If yes, details…………………………..
(8) Place of purchase ………………Place of sale ………… Income…………………
(9) Have you started keeping new breeds? Yes/No
If yes, name the breed ……………….. Profitability ………………
(10) Major problems in it …………………………………………………..
(11) Any suggestions for improvement ………………………………

II. *Piggeries*

(1) Do your practice piggeries? Yes/No, If yes, since when ....
(2) Commercial/subsistence.........
(3) Main source/subsidiary source
(4) Average income per month from it ....
(5) Where and how do you sell its products ....
(6) Do you rear new breeds – Yes/No
Name .... Profitability ....
(7) Do you face any social problem in practicing it? Yes/No
(8) Major problems ....
(9) Suggestions ....

III. *Fishing*

(1) Do you practice fishing? Yes/No, If yes, since when? ....
(2) Commercial/subsistence basis........
(3) Main Source/Subsidiary source
(4) Do you have a natural fishing pond? Yes/No, If yes, do you have a sole possession of it or do you get it on rental basis from the government? ....
(5) If it is on rental basis, what is the nature of the contract? ....
How much money do you pay? ....
(6) Do you have a man-made pond? Yes/No, If yes, year of digging .... Area ....Amount of fish catch ....
(7) Where do you sell fish? ....
(8) Peak season of fishing: Summer/Winter/Rainy season
(9) Do you purchase and sell fishes? Give details
....
....
(10) Major problems ....
(11) Suggestions
....

IV. *Poultry Farming*

(1) Do you practice poultry farming? Yes/No, If yes, since when? ....
(2) Commercial/subsistence basis (3)Main source/subsidiary source

(4) Do you practice at home/poultry farm? Yes/No ....
(5) If at home,
  (a) how many birds? ....
  (b) How much do you earn monthly? ....
(6) If at poultry farm,
  (a) When did you construct it? ....
  (b) What is its size and capacity? ....
  (c) How many birds do you normally keep at a time? ....
(7) How and where do you sell its products? ....
(8) How much do you spend on an average per month on it? ....
(9) How much do you get on an average per month from it? ....
(10) What are the new breeds you have? ....
  Profitability ....
(11) Problems
  ....
(12) Suggestions
  ....

V. Apiculture

(1) Do you practice bee farming? Yes/No, If yes, since when? ....
(2) Where do you practice it? ....
(3) Do you have a bee farm? Yes/No, If yes, (a) what is its size? ....
(b) How many boxes do you have? ....
(4) How much is the average production per month
(5) How much honey do you get on an average in a month? ....
(6) Where and how do you sell it? ....
(7) What is its future prospects in your area? ....
(8) Major problems ....
(9) Suggestions ....

VI. Vegetable Farming

(1) Do you grow vegetables on commercial basis? Yes/No
(2) If yes, how much area do you devote to its cultivation? ....
(3) Pattern of landuse during—

Kharif Season

Vegetables	*Area*	*Total Expenses*	*Output*	*Place of sale*	*Return per unit*	*Remarks*

Rabi Season

Vegetables	*Area*	*Total Expenses*	*Output*	*Place of sale*	*Return per unit*	*Remarks*

Zaid Season

Vegetables	*Area*	*Total Expenses*	*Output*	*Place of sale*	*Return per unit*	*Remarks*

(4) Do you also purchase and sell vegetables? Yes/No
If yes,
(a) Place of Purchase………………
(b) Place of sale ……………………
(c) Profit per month…………………
(5) Major Problems ……………………………………………………
(6) Suggestions ………………………………………………………

VII. Floriculture

(1) Do you practice floriculture on a commercial basis? Yes/No
(2) If yes, how much area do you devote to flower cultivation? ……………………
(3) Pattern of landuse during –

Kharif Season

Flowers	*Seasonal/ Perennial*	*Area*	*Expenses*	*Mode of sale*	*Place of Sale*	*Income*	*Remarks*

Rabi Season

Flowers	*Seasonal/ Perennial*	*Area*	*Expenses*	*Mode of sale*	*Place of Sale*	*Income*	*Remarks*
....	....	....	....	....	....	....	....
....	....	....	....	....	....		
....	....	....	....	....	....	....	....

Zaid Season

Flowers	*Seasonal/ Perennial*	*Area*	*Expenses*	*Mode of sale*	*Place of Sale*	*Income*	*Remarks*
	....	....	....	....	....	....	....
	....	......	....	....	....	....	....

(4) Major problems ....
(5) Suggestions ....

VIII. Horticulture (Fruits and Medicinal Plant)

(1) Do you grow fruits on a commercial basis? Yes/No
(2) If yes, how much area do you devote to its cultivation? ....

(3)	*Fruit*	*No. of Plants*	*Area*	*Fruit bearing season*	*Amount of fruit per season*	*Expenses*	*Income*

(4) Future prospects ....
(5) Problems ....
(6) Suggestions ....

IX. Purchase of Edibles

1. Do you purchase foodgrains for home consumption? If yes, give the following details.

Commodities	*Monthly Quantity*	*Expenses*

2. From where do you purchase? PDS/Open market
3. What are the other edibles which you purchase?...........................

Edibles	*Monthly Quantity*	*Expenses*

X. Sale of Edibles

1. Do you sell agricultural commodities you produce? Yes/No, If yes

Commodities	*Yearly/Monthly Quantity*	*Expenses*

2. Where do you sell? Within village/out side village.

XI. Personal Observation of Interviewer:

....

APPENDIX-II

Questionnaire for Village Survey

1. **Name of village** ....
   Tahsil ....
   District..
2. **Population-**
   Total ........................M.........................F..................................
   Hindu ....Muslim.....................Others...........................
3. **Total Area:** ....
   Cultivated area ....
   Forest area...
4. **Major Crops:**

	At present		10 year before	
	Area	Yield	Area	Yield
i. Rice		....	....	....
ii. Wheat		....	....	....
iii.		....	....	....
iv.		....	....	....
v.		....	....	....

5. **Agricultural Mechanization:**

Farm implants		Chemical ferti.		HYV		Irrigation	
Name	No.	Name	Quantity	Name	Quantity	Source	Area
....	....	......	....	......	....	....	......
....	....	......	....	......	....	....	......
....	....	......	....	......	....	....	......
....	....	......	....	......	....	....	......

6. **Dairy Farming:**
   No. of people............... Comm./Subsi. .... Type & No. of Animals .... Total Income.......................
   Major Problems ....
7. **Piggeries:**
   No. of people .... Comm./Subsi. .... Type & No. of Animals .... Total Income ....
   Major Problems ....
8. **Fishing:**
   No. of ponds .... Comm./Subsi. .... Season of fishing .... Total Income.................................
   Major Problems ....
9. **Poultry Farming:**
   No. of Farms............... Comm./Subsi. .... Type & No. of Birds .... Total Income ....
   Major Problems ....
10. **Apiculture:**
    Total area .... Comm./Subsi. .... Total Income.................................
    Major Problems ....
11. **Vegetable Farming:**
    Total area .... Comm./Subsi. .... Total Income ....
    Major Vegetables ....
    Major Problems ....
12. **Floriculture:**
    Total area .... Comm./Subsi. .... Total Income ....
    Major Flowers ....
    Major Problems ....
13. **Horticulture:**
    Total area .... Comm./Subsi. .... Total Income..........................
    Major Fruits ....
    Major Problems ....
14. **Herbal Cultivation:**
    Total area.............. Comm./Subsi. ....
    Total Income .... Major Herbal Plants ....
    Major Problems ....

Signature with Date

APPENDIX-III

Uttarakhand : Climatological Table of Observatories- 1951-1980

A: Mean Maximum and Minimum Temperature

Station	Temperature (°c)																							
	January		February		March		April		May		June		July		August		September		October		November		December	
	Mean Max.	Mea n Min	Mean Max.	Mea n Min.	Mean Max.	Mean Min.	Mean Max.	Mean Min.	Mean Max.	Mean Min.	Mean Max.	Mean Min.	Mean Max.	Mean Min.	Mean Max.	Mean Min.	Mean Max.	Mean Min.	Mean Max.	Mean Min.	Mean Max.	Mean Min.	Mean Max.	Mean Min.
1. Dehradun	19.4	5.8	21.9	7.9	26.7	12.2	32.3	16.9	35.8	20.5	34.9	23.0	30.5	22.5	29.5	22.0	29.7	20.6	28.5	15.9	25.1	10.4	21.3	6.8
2. Joshimath	11.7	2.0	13.0	3.3	17.9	6.4	22.5	10.9	25.4	14.0	26.4	16.6	24.5	16.9	23.8	16.8	23.2	14.8	21.1	10.4	17.0	6.3	13.8	3.9
3. Mukhim	12.5	3.5	13.5	4.4	18.0	7.9	22.7	12.0	25.1	15.0	25.9	16.8	23.8	17.1	23.4	16.9	23.5	15.3	21.9	11.8	18.3	7.9	14.9	5.3
4. Mukteshwer	10.7	1.5	12.5	2.8	16.7	6.2	21.3	10.4	23.9	12.9	23.5	14.3	20.9	14.2	20.4	14.0	20.3	12.5	19.3	9.5	16.4	6.0	13.5	3.5
5. Mussorri	10.6	2.5	12.4	3.8	16.5	7.5	21.2	12.0	24.1	14.7	23.7	16.2	20.8	15.5	20.2	15.2	19.9	14.0	18.9	11.2	15.9	7.3	13.1	4.6
6. Nainital	10.7	1.7	12.3	3.5	16.3	7.5	20.0	11.9	23.5	14.6	23.5	16.4	26.6	16.5	21.0	16.0	20.7	14.1	18.7	9.7	15.4	5.7	12.9	3.1
7. Roorkee	20.4	6.1	23.5	8.2	29.1	12.8	35.7	18.3	39.2	22.1	38.1	24.9	33.5	24.8	32.4	24.5	32.7	22.9	31.4	17.2	27.1	10.5	22.3	6.7
8. Tehri	19.6	4.6	23.3	7.1	27.6	11.0	33.0	15.6	36.3	19.0	36.5	22.8	33.6	23.4	32.5	23.2	32.2	20.9	30.5	15.4	26.0	9.3	21.6	4.8

(Contd.)

Appendix III (*Contd.*)

B: Mean Monthly Rainfall

Station	Rainfall in mm.												Annual Rainfall	Average Rainfall
	January	February	March	April	May	June	July	August	September	October	November	December		
1. Askot	49.30	57.80	60.70	51.30	92.40	436.20	700.60	692.30	359.80	57.90	7.40	20.20	2585.90	215.49
2. Dehradun	51.60	48.10	51.30	17.30	36.20	225.00	718.60	734.00	321.70	47.50	11.30	23.20	2285.80	190.48
3. Dharchula	48	57.9	66.7	37.4	79.9	321.4	604.3	570.5	325.1	64	12.2	19.9	2207.3	183.94
4. Joshimath	77.80	107.60	113.00	59.10	52.90	111.50	267.10	232.70	124.30	40.40	14.90	24.30	1225.60	102.13
5. Mukhim	62.20	70.40	76.30	46.70	68.30	167.60	416.30	405.70	212.20	41.60	11.80	30.00	1609.10	134.09
6. Mukteshwer	54.50	46.40	55.70	27.60	58.00	131.60	310.60	298.80	209.70	78.70	6.30	18.60	1296.50	108.04
7. Munsyari	189.50	117.80	63.60	47.80	22.80	18.20	75.40	73.40	124.70	11.90	7.80	25.40	778.30	64.85
8. Mussoorri	51.80	52.80	57.70	30.00	58.40	174.90	662.00	670.60	277.50	64.70	14.80	18.20	2133.40	177.78
9. Nainital	81.10	60.00	54.40	32.20	76.90	331.80	759.30	587.10	348.90	150.70	6.90	26.90	2516.20	209.68
10. Roorkee	36.90	33.00	34.80	8.40	19.20	128.70	342.60	336.80	157.80	38.40	5.50	14.30	1156.40	96.36
11. Tehri	49.8	53.7	54.4	29.3	43.1	79.7	220.8	230.4	126.6	36.1	9.2	28.9	962	80.16

Source: Climatological Tables of Observatories in india, 1951–1980, Director General, Indian Metrological Department, New Delhi, Fifth Edition - 1999.

APPENDIX-IV

Demographic Characteristics of Uttarakhand : 2001

District	Total Population (2001)	% to Total Pop.	Rural Pop.	Urban Pop.	Male	Female	Growth rate	Pop. Density		Literacy rate	
							2001-11	2001	2011	2001	2011
(1)	(2)	(3)	(4)	(5)	(6)	(7)	(8)	(9)	(10)	(11)	(12)
1. Chamoli	370359	4.36	319656 86.31%	50703 13.59%	183745 49.61%	186614 50.39%	5.6	46	49	75.4	83.48
2. Dehradun	1282143	15.10	603401 47.06%	678742 52.94%	679583 53.00%	602560 47.00%	32.48	415	550	79	85.24
3. Hardwar	1447187	17.05	1000912 69.1 6%	446275 30.84%	776021 53.62%	671166 46.38%	33.16	613	817	63.7	74.62
4. Pauri Garhwal	697078	8.21	607203 87.11%	89875 12.89%	331061 47.49%	366017 52.51%	–1.51	131	129	77.5	82.59
5. Rudraprayag	227439	2.68	224707 98.80%	2732 1.20%	107535 47.28%	119904 52.72%	4.14	115	119	73.6	82.09
6. Tehri Garhwal	604747	7.12	544901 90.10%	59846 9.90%	295168 48.81%	309579 51.19%	1.93	166	169	66.7	75.1
7. Uttarkashi	295013	3.48	272095 92.23%	22918 7.77%	152016 51 .53%	142997 48.47%	11.75	37	41	65.7	75.98
Garhwal Division	**4923966**	**58.00**	**3572875 72.56%**	**1351091 27.44%**	**2525129 51 .28%**	**2398837 48.72%**	—	—	—		—

(Contd.)

Appended III (*Contd.*)

District	*Total Population (2001)*	*% to Total Pop.*	*Rural Pop.*	*Urban Pop.*	*Male*	*Female*	*Growth rate*	*Pop. Density*		*Literacy rate*	
							2001-11	2001	2011	2001	2011
(1)	(2)	(3)	(4)	(5)	(6)	(7)	(8)	(9)	(10)	(11)	(12)
8. Almora	630567	7.43	576062 91 .36%	54505 8.64%	293848 46.60%	336719 53.40%	–1.73	201	198	73.6	81.06
9. Bageshwar	249462	2.94	241659 96.87%	7803 3.13%	118510 47.51%	130952 52.49%	5.13	110	116	71.3	80.69
10. Champawat	224542	2.64	190764 84.96%	33778 15.04%	111084 49.47%	113458 50.53%	15.49	127	147	70.4	80.73
11. Nainital	762909	8.99	493859 64.73%	269050 35.27%	400254 52.46%	362655 47.54%	25.2	179	225	78.4	84.85
12. Pithoragarh	462289	5.45	402456 87.06%	59833 12.94%	227615 49.24%	234674 50.76%	5.13	65	69	75.9	82.93
13. US Nagar	1235614	14.55	832600 67.38%	403014 32.62%	649484 52.56%	586130 47.44%	33.4	486	648	64.9	74.44
Kumaon Division	**3565383**	**42.00**	**2737400 76.78%**	**827983 23.22%**	**1800795 50.51%**	**1764588 49.49%**	—	—	—		—
UTTARAKHAND	**8489349**	**100.00**	**6310275 74.33%**	**2179074 25.67%**	**4325924 50.96%**	**4163425 49.04%**	**19.17**	**613**	**817**	**76.6**	**79.63**

APPENDIX-V

Patterns of General Landuse in Uttarakhand : 1992-1993

District*	Land use Categories (Area in Hec.)											G.C.A.
	1. Forests	%age of Total Reporting area	2.Not available for cultivation	%age of Total Reporting area	3.Other uncultivated land excluding fallow land	%age of Total Reporting area	4.Fallow land	%age of Total Reporting area	5.Net sown area	%age of Total Reporting area	6. Total Reporting area	
1. Chamoli	520361	61.90	182669	21.73	90362	10.75	1701	0.20	45611	5.43	840704	70092
2. Dehradun	219812	69.66	18724	5.93	16314	5.17	7449	2.36	53255	16.88	315554	86488
3. Pauri Garhwal	450323	59.35	52940	6.98	152734	20.13	18405	2.43	84308	11.11	758710	130999
4. Tehri Garhwal	397201	69.13	22933	3.99	74706	13.00	8338	1.45	71336	12.42	574544	113699
5. Uttarkashi	710278	88.60	27042	3.37	30659	3.82	3915	0.49	29725	3.71	801691	44669
Garhwal Division	**2297975**	**69.82**	**304308**	**9.25**	**364775**	**11.08**	**39808**	**1.21**	**404931**	**12.30**	**3291203**	**626799**

(Contd.)

Appendix V (*Contd.*)

District	*Land use Categories (Area in Hec.)*												G.C.A.
	1. Forests	*%age of Total Reporting area*	*2.Not available for cultivation*	*%age of Total Reporting area*	*3.Other uncultivated land excluding fallow land*	*%age of Total Reporting area*	*4.Fallow land*	*%age of Total Reporting area*	*5.Net sown area*	*%age of Total Reporting area*	*6. Total Reporting area*		
6. Almora	392148	53.95	50183	6.90	169231	23.28	8558	1.18	106760	14.69	726880		178593
7. Nainital	406160	57.71	37883	5.38	48908	6.95	10576	1.50	200216	28.45	703743		340870
8. Pithoragarh	330243	51.84	41274	6.48	180420	28.32	13769	2.16	71387	11.21	637093		125965
Kumaon Division	**1128551**	**54.58**	**129340**	**6.26**	**398559**	**19.28**	**32903**	**1.59**	**378363**	**18.30**	**2067716**		**645428**
UTTARAKHAND	**3426526**	**63.94**	**433648**	**8.09**	**763344**	**14.24**	**72711**	**1.36**	**783294**	**14.62**	**5358919**		**1272227**

Note : *There were only 8 districts in Uttarakhand in 1992-93.

APPENDIX-VI

Uttarakhand : District-wise Patterns of General Landuse 2005-06

(Area : in hec.)

			Not Available for Cult.			Area under misc.uses			
1. District	*2. Reporting area*	*3. Forests*	*4. Area under non-agri .uses*	*5. Barren & uncultura-ble land*	*6. Total (4&5)*	*7. Permanent pastures & other grazing lands*	*8. Land under misc. tree crops and groves not including in net area*	*9. Culturable waste land*	*10. Total (7,8,& 9)*
1. Chamoli	847580	506100 59.71%	8221 0.97%	158880 18.75%	167101 19.72%	48615 5.74%	40600 4.79%	49808 5.88%	139023 16.40%
2. Dehradun	368944	201831 54.71%	21589 5.85%	3556 0.96%	25145 6.82%	289 0.08%	15385 4.17%	64138 17.38%	79812 21.63%
3. Hardwar	231076	72431 31.35%	26944 11.66%	2416 1.05%	29360 12.71%	63 0.03%	738 0.32%	2023 0.88%	2824 1.22%
4. Pauri Garhwal	673431	385099 57.18%	15794 2.35%	36006 5.35%	51800 7.69%	35130 5.22%	55771 8.28%	38128 5.66%	129029 19.16%
5. Rudraprayag	235421	179895 76.41%	3084 1.31%	7324 3.11%	10408 4.42%	8158 3.47%	11600 4.93%	4378 1.86%	24136 10.25%
6. Tehri Garhwal	485766	322051 66.30%	5354 1.10%	6005 1.24%	11359 2.34%	593 0.12%	146 0.03%	78165 16.09%	78904 16.24%
7. Uttarkashi	812414	721661 88.83%	5189 0.64%	37572 4.62%	42761 5.26%	8267 1.02%	4450 0.55%	2278 0.28%	14995 1.85%
Garhwal Division	**3654632**	**2389068** **65.37%**	**86175** **2.36%**	**251759** **6.89%**	**337934** **9.25%**	**101115** **2.77%**	**128690** **3.52%**	**238918** **6.54%**	**468723** **12.83%**

(Contd.)

Appendix VI *(Contd.)*

			Not Available for Cult.			Area under misc.uses			
1. District	*2.Reporting area*	*3.Forests*	*4.Area under non-gri .uses*	*5.Barren & uncultura-ble land*	*6.Total (4&5)*	*7.Perman-ent pastures & other grazing lands*	*8.Land under misc.tree crops and groves not including in net area*	*9.Cultura ble waste land*	*10. Total (7,8,& 9)*
8. Almora	465858	236179	12655	25647	38302	30511	27523	42461	100495
		50.70%	2.72%	5.51%	8.22%	6.55%	5.91%	9.11%	21.57%
9. Bageshwar	213542	110160	4697	6779	11476	27346	20333	16774	64453
		51.59%	2.20%	3.17%	5.37%	12.81%	9.52%	7.86%	30.18%
10. Champawat	235388	132337	4571	5302	9873	18378	26043	15013	59434
		56.22%	1.94%	2.25%	4.19%	7.81%	11.06%	6.38%	25.25%
11. Nainital	406475	298336	9555	1272	10827	3	17954	26908	44865
		73.40%	2.35%	0.31%	2.66%	0.00%	4.42%	6.62%	11.04%
12. Pithoragarh	410638	205239	9979	20824	30803	53026	26990	40798	120814
		49.98%	2.43%	5.07%	7.50%	12.91%	6.57%	9.94%	29.42%
13. US Nagar	279234	93738	26461	998	27459	26	1401	2740	4167
		33.57%	9.48%	0.36%	9.83%	0.01%	0.50%	0.98%	1.49%
Kuman Division	**2011135**	**1075989**	**67918**	**60822**	**128740**	**129290**	**120244**	**144694**	**394228**
		53.50%	**3.38%**	**3.02%**	**6.40%**	**6.43%**	**5.98%**	**7.19%**	**19.60%**
UTTARAKHAND	**5665767**	**3465057**	**154093**	**312581**	**466674**	**230405**	**248934**	**383612**	**862951**
		61.16%	**2.72%**	**5.52%**	**8.24%**	**4.07%**	**4.39%**	**6.77%**	**15.23%**

(Contd.)

Appendix VI (*Contd.*)

1. District	*Fellow Land*			*14. NSA (Net area sown)*	*15. Total Cropped area*	*16. Cropping intensity*
	11. Fallow land other then current fallows	*12. Current fallow*	*13. Total (11&12)*			
1. Chamoli	1105	1691	2796	32560	45440	139.56
	0.13%	0.20%	0.33%	3.84%	5.36%	
2. Dehradun	7016	6632	13648	48508	73679	151.89
	1.90%	1.80%	3.70%	13.15%	19.97%	
3. Hardwar	3529	2351	5880	120581	168066	139.38
	1.53%	1.02%	2.54%	52.18%	72.73%	
4. Pauri Garhwal	18297	7715	26012	81491	124674	152.99
	2.72%	1.15%	3.86%	12.10%	18.51%	
5. Rudraparyag	680	786	1466	19516	27999	143.47
	0.29%	0.33%	0.62%	8.29%	11.89%	
6. Tehri Garhwal	6276	8056	14332	59120	90946	153.83
	1.29%	1.66%	2.95%	12.17%	18.72%	
7. Uttarkashi	1615	563	2178	30819	47491	154.10
	0.20%	0.07%	0.27%	3.79%	5.85%	
Garhwal Division	**38518**	**27794**	**66312**	**392595**	**578295**	**147.30**
	1.05%	**0.76%**	**1.81%**	**10.74%**	**15.82%**	

(*Contd.*)

Appendix VI (*Contd.*)

1. District	*Fellow Land*			*14. NSA (Net area sown)*	*15. Total Cropped area*	*16. Cropping intensity*
	11. Fallow land other then current fallows	*12. Current fallow*	*13. Total (11&12)*			
8. Almora	8403	1528	9931	80951	124938	154.34
	1.80%	0.33%	2.13%	17.38%	26.82%	
9. Bageshwar	533	353	886	26567	47576	179.08
	0.25%	0.17%	0.41%	12.44%	22.28%	
10. Champawat	4904	4875	9779	23965	39295	163.97
	2.08%	2.07%	4.15%	10.18%	16.69%	
11. Nainital	2699	2018	4717	47730	81733	171.24
	0.66%	0.50%	1.16%	11.74%	20.11%	
12. Pithoragarh	4013	3867	7880	45902	79992	174.27
	0.98%	0.94%	1.92%	11.18%	19.48%	
13. USNagar	2535	1474	4009	149861	260538	173.85
	0.91%	0.53%	1.44%	53.67%	93.30%	
Kumaun Division	**23087**	**14115**	**37202**	**374976**	**634072**	**169.1**
	1.15%	**0.70%**	**1.85%**	**18.64%**	**31.53%**	
UTTARAKHAND	**61605**	**41909**	**103514**	**767571**	**1212367**	**157.95**
	1.09%	**0.74%**	**1.83%**	**13.55%**	**21.40%**	

APPENDIX-VII

Uttarakhand : District-wise Sources of Irrigation : 2005-06

(Area : In Hec.)

District	Sources of Irrigation									
	Canals		*Wells*		*Tank*		*Other sources*		*Total*	
	Area	% to Total	Area	% to Total	Area	% to Total	Area	% to Total	Area	% to Total
1. Chamoli	841	45.91	0	0.00	0	0.00	991	54.09	1832	0.53
2. Dehradun	13997	62.24	3309	14.72	0	0.00	5181	23.04	22487	6.55
3. Hardwar	15080	14.05	90209	84.03	0	0.00	2064	1.92	107353	31.26
4. Pauri Garhwal	3037	37.78	476	5.92	21	0.26	4504	56.03	8038	2.34
5. Rudraparyag	1403	64.51	0	0.00	0	0.00	772	35.49	2175	0.63
6. Tehri Garhwal	778	10.12	0	0.00	0	0.00	6913	89.88	7691	2.24
7. Uttarkashi	3468	69.72	0	0.00	0	0.00	1506	30.28	4974	1.45
Garhwal Division	**38604**	**24.98**	**93994**	**60.82**	**21**	**0.01**	**21931**	**14.19**	**154550**	**45.00**

(Contd.)

Appendix VII (*Contd.*)

District	*Sources of Irrigation*									
	Canals		*Wells*		*Tank*		*Other sources*		*Total*	
	Area	% to Total	Area	% to Total	Area	% to Total	Area	% to Total	Area	% to Total
8. Almora	3178	63.33	0	0.00	0	0.00	1840	36.67	5018	1.46
9. Bageshwar	4633	79.76	0	0.00	0	0.00	1176	20.24	5809	1.69
10. Champawat	1325	60.86	655	30.09	0	0.00	197	9.05	2177	0.63
11. Nainital	21363	81.88	4577	17.54	0	0.00	152	0.58	26092	7.60
12. Pithoragarh	1255	30.10	0	0.00	0	0.00	2915	69.90	4170	1.21
13. US Nagar	25250	17.34	114354	78.54	313	0.21	5679	3.90	145596	42.40
Kumaon Division	**57004**	**30.18**	**119586**	**63.32**	**313**	**0.17**	**11959**	**6.33**	**188862**	**55.00**
UTTARAKHAND	**95608**	**27.84**	**213580**	**62.19**	**334**	**0.10**	**33890**	**9.87**	**343412**	**100.00**

APPENDIX-VIII

District-wise Pattern of Agricultural Landuse : 2005-06

(Area: in hec.)

District	1. Rice	X	2. Maize	X	3. Madua	X	4. Wheat	X	5. Barley	X	6. Pulses	X	7. Sugar	X
1. Chamoli	10219	22.49	195	0.43	8665	19.07	13492	29.69	1547	3.40	1635	3.60	0	0.00
2. Dehradun	12344	16.75	10917	14.82	2692	3.65	23057	31.29	1052	1.43	3522	4.78	4899	6.65
3. Hardwar	17137	10.20	1335	0.79	0	0.00	42147	25.08	23	0.01	1564	0.93	70591	42.00
4. Pauri Garhwal	22335	17.91	2305	1.85	28363	22.75	35005	28.08	6777	5.44	7259	5.82	10	0.01
5. Rudraparyag	8016	28.63	194	0.69	5719	20.43	9054	32.34	1056	3.77	683	2.44	0	0.00
6. Tehri Garhwal	12164	13.37	1972	2.17	13733	15.10	28409	31.24	2297	2.53	7783	8.56	0	0.00
7. Uttarkashi	10211	21.50	681	1.43	7022	14.79	13413	28.24	305	0.64	4158	8.76	0	0.00
Garhwal Division	**92426**	**15.98**	**17599**	**3.04**	**66194**	**11.45**	**164577**	**28.46**	**13057**	**2.26**	**26604**	**4.60**	**75500**	**13.06**
8. Almora	21212	16.98	1887	1.51	33648	26.93	39917	31.95	2837	2.27	5597	4.48	0	0.00
9. Bageshwar	15699	33.00	1468	3.09	5840	12.28	17565	36.92	1786	3.75	2258	4.75	0	0.00
10. Champawat	9395	23.91	853	2.17	6655	16.94	11440	29.11	1393	3.54	3208	8.16	7	0.02
11. Nainital	12378	15.14	5830	7.13	3481	4.26	26205	32.06	940	1.15	2969	3.63	4550	5.57
12. Pithoragarh	22696	28.37	3585	4.48	8511	10.64	26082	32.61	3382	4.23	7289	9.11	0	0.00
13. US Nagar	110624	42.46	239	0.09	50	0.02	93451	35.87	25	0.01	3635	1.40	27691	10.63
Kumaon Division	**192004**	**30.28**	**13862**	**2.19**	**58185**	**9.18**	**214660**	**33.85**	**10363**	**1.63**	**24956**	**3.94**	**32248**	**5.09**
UTTARAKHAND	**284430**	**23.46**	**31461**	**2.60**	**124379**	**10.26**	**379237**	**31.28**	**23420**	**1.93**	**51560**	**4.25**	**107748**	**8.89**

(Contd.)

Appendix-VIII (*Contd.*)

	8. Condiments	*X*	*9. Fruits*	*X*	*10. Vegeta.*	*X*	*11. Oil seeds*	*X*	*12. Drugs, Narco. & Plantation*	*X*	*15. Total cropped area*
1. Chamoli	241	0.53	13	0.03	2201	4.84	986	2.17	23	0.05	45440
2. Dehradun	1198	1.63	3069	4.17	3493	4.74	1529	2.08	641	0.87	73679
3. Hardwar	26	0.02	6092	3.62	4092	2.43	2500	1.49	52	0.03	168066
4. Pauri Garhwal	747	0.60	84	0.07	1365	1.09	1033	0.83	0	0.00	124674
5. Rudraparyag	141	0.50	10	0.04	283	1.01	266	0.95	0	0.00	27999
6. Tehri Garhwal	638	0.70	5	0.01	2362	2.60	2137	2.35	12	0.01	90946
7. Uttarkashi	110	0.23	504	1.06	3836	8.08	2365	4.98	1	0.00	47491
Garhwal Division	**3101**	**0.54**	**9777**	**1.69**	**17632**	**3.05**	**10816**	**1.87**	**729**	**0.13**	**578295**
8. Almora	2165	1.73	101	0.08	2184	1.75	842	0.67	2	0.00	124938
9. Bageshwar	383	0.81	0	0.00	719	1.51	203	0.43	58	0.12	47576
10. Champawat	1353	3.44	54	0.14	1515	3.86	1380	3.51	2	0.01	39295
11. Nainital	1255	1.54	4550	5.57	7775	9.51	9125	11.16	49	0.06	81733
12. Pithoragarh	666	0.83	0	0.00	2686	3.36	2288	2.86	0	0.00	79992
13. US Nagar	92	0.04	1414	0.54	2291	0.88	9389	3.60	83	0.03	260538
Kumaon Division	**5914**	**0.93**	**6119**	**0.97**	**17170**	**2.71**	**23227**	**3.66**	**194**	**0.03**	**634072**
UTTARAKHAND	**9015**	**0.74**	**15896**	**1.31**	**34802**	**2.87**	**34043**	**2.81**	**923**	**0.08**	**1212367**

X: Percentage to net shown Area

APPENDIX- IX

Division-wise Area Under Principal Crops : 1992-93, 2003-04, 2005-06

(Area : in Hect.)

Crops	1992-93			2003-04			2005-06		
	Area	% to N.S.A.	% to G.C.A.	Area	% to N.S.A.	% to G.C.A.	Area	% to N.S.A.	% to G.C.A.
1. Rice									
Garhwal Div.	99986	24.69	15.95	102028	25.66	16.90	92426	23.54	15.98
Kumaon Div.	175260	46.32	27.15	175298	48.26	28.36	192004	51.20	30.28
Uttarakhand	275246	35.13	21.63	277326	36.45	22.70	284430	37.05	23.46
2. Maize									
Garhwal Div.	22488	5.55	3.58	16804	4.22	2.78	17599	4.48	3.04
Kumaon Div.	10577	2.79	1.63	14542	4.00	2.35	13862	3.69	2.18
Uttarakhand	33065	4.22	2.59	31346	4.12	2.56	31461	4.09	2.59
3. Mandua									
Garhwal Div.	74197	18.32	11.87	70470	17.72	11.67	66194	16.86	11.44
Kumaon Div.	65118	17.21	10.08	55953	15.40	9.05	58185	15.51	9.17
Uttarakhand	139315	17.78	10.95	126423	16.61	10.34	124379	16.20	10.25
4. Wheat									
Garhwal Div.	186307	46.00	29.72	172741	43.45	28.62	164577	41.92	28.45

(Contd.)

Appendix IX (*Contd.*)

Crops	*1992-93*			*2003-04*			*2005-06*		
	Area	*% to N.S.A.*	*% to G.C.A.*	*Area*	*% to N.S.A.*	*% to G.C.A.*	*Area*	*% to N.S.A.*	*% to G.C.A.*
Kumaon Div.	222730	58.86	34.50	210645	57.99	34.08	214660	57.24	33.85
Uttarakhand	409037	52.22	32.15	383386	50.39	31.38	379237	49.40	31.28
5. Barley									
Garhwal Div.	14622	3.61	2.33	13499	3.39	2.23	13057	3.32	2.25
Kumaon Div.	13148	3.47	2.03	11441	3.14	1.85	10363	2.76	1.63
Uttarakhand	27770	3.54	2.18	24940	3.27	2.04	23420	3.05	1.93
6. Pulses									
Garhwal Div.	27834	6.87	4.44	26734	6.72	4.42	26604	6.77	4.60
Kumaon Div.	24368	6.44	3.77	23834	6.56	3.85	24956	6.65	3.93
Uttarakhand	52202	6.66	4.10	50568	6.64	4.13	51560	6.71	4.25
7. Sugarcane									
Garhwal Div.	62357	15.39	9.94	78076	19.64	12.93	75500	19.23	13.05
Kumaon Div.	44939	11.87	6.96	38801	10.68	6.27	32248	8.60	5.08
Uttarakhand	107296	13.69	8.43	116877	15.36	9.56	107748	14.03	8.88
8. Vegeta.									
Garhwal Div.	11788	2.91	1.88	16706	4.20	2.76	17632	4.49	3.04
Kumaon Div.	11190	2.95	1.73	16441	4.52	2.66	17170	4.57	2.70
Uttarakhand	22978	2.93	1.80	33147	4.35	2.71	34802	4.53	2.87
9. Oilseeds									
Garhwal Div.	12698	3.13	2.02	9600	2.41	1.59	10816	2.75	1.87
Kumaon Div.	24454	6.46	3.78	20874	5.74	3.37	23227	6.19	3.66
UTTARAKHAND	37152	4.74	2.92	30474	4.00	2.49	34043	4.43	2.80

APPENDIX-X

Crop Combination Analysis

(1)		(2)		(3)	(4)	(5)	(6)	(7)	(8)	(9)	(10)	(11)	(12)	(13)	(14)
A	*B*	*A*	*B*	*A*	*B*	*B*	*B*	*B*	*B*	*B*	*B*	*B*	*B*	*B*	*B*
70.5	28.79	48.5	20.64	94		...	...	...	...	...	...	...	...	...	6.51
70.0	27.64	78.0	20.00	93	...	...	...	...	...	...	...	6.97	6.27	5.68	5.20
69.5	26.55	77.5	19.58	92	...	...	...	...	...	...	7.42	6.59	5.92	5.39	4.94
69.0	25.51	77.0	18.81	91	...	....	...	...	...	8.12	7.02	6.25	5.63	5.13	4.68
68.5	24.50	76.5	18.25	90	...	...	...	...	8.84	7.60	6.67	5.94	5.35	4.88	4.49
68.0	**23.54**	**76.0**	**17.71**	**89**	**...**	**...**	**...**	**10.00**	**8.38**	**7.22**	**6.35**	**5.66**	**5.11**	**4.67**	**4.29**
67.5	22.61	75.5	17.03	88	...	...	11.63	9.47	7.97	6.89	6.06	5.41	4.89	4.47	4.10
67.0	21.72	75.0	16.67	87	...	...	11.09	9.00	7.59	6.57	5.78	5.17	4.68	4.28	3.92
66.5	20.85	...	...	86	...	13.68	10.52	8.57	7.24	6.27	5.52	4.94	4.48	4.09	3.76
66.0	20.00	74.5	16.03	85	...	12.93	10.00	88.16	6.92	5.99	5.28	4.73	4.29	3.91	3.60
65.5	**19.18**	**74.0**	**15.68**	**84**	**...**	**12.25**	**9.52**	**7.78**	**6.60**	**5.73**	**5.05**	**4.53**	**4.10**	**3.75**	**3,45**
65.0	18.38	73.5	15.21	83	16.34	11.63	9.07	7.43	6.30	5.47	4.83	4.33	3.93	3.60	3.20
...	...	73.0	14.76	82	15.44	11.06	8.64	7.09	6.02	5.23	4.62	4.14	3.76	3.45	3.16
64.5	17.60	72.5	14.33	81	14.58	10.61	3.24	6.77	5.75	5.00	4.42	3.96	3.60	3.29	3.03

(Contd.)

Appendix X (*Contd.*)

(1)		(2)		(3)	(4)	(5)	(6)	(7)	(8)	(9)	(10)	(11)	(12)	(13)	(14)
64.0	16.83	72.0	13.90	80	13.83	10.00	7.85	6.45	5.49	4.77	4.22	3.79	3.35	3.13	2.90
63.5	**16.09**	**.71.5**	**13.48**	**79**	**13.10**	**9.51**	**7.48**	**6.17**	**5.24**	**4.56**	**4.03**	**3.62**	**3.29**	**3.01**	**2.77**
63.0	15.39	71.0	13.08	78	12.43	9.05	7.12	5.88	5.00	4.35	3.85	3.45	3.14	2.88	2.65
62.5	14.64	705	12.66	77	11.78	8.60	6.73	5.60	4.77	4.15	3.68	3.30	3.00	2.75	2.53
62.0	13.94	70.0	12.25	76	11.16	8.17	6.45	5.32	4.45	3.96	3.50	3.24	2.85	2.62	2.41
61.5	13.26	...	...	75	10.57	7.75	6.13	5.06	4.32	3.76	3.33	2.99	2.71	2.43	2.29
61.0	**12.58**	**69.5**	**11.90**	**74**	**10.00**	**7.35**	**5.32**	**4.82**	**4.10**	**3.58**	**3.17**	**2.84**	**2.58**	**2.37**	**2.18**
60.5	11.92	69.0	11.50	73	9.46	6.96	5.51	4.55	3.89	3.40	3.00	2.70	2.45	2.26	2.07
60.0	11.27	68.5	11.10	72	8.93	6.58	5.22	4.32	3.69	3.21	2.85	2.56	2.33	2.14	1.97
...	...	68.0	10.73	71	8.42	6.22	4.93	4.07	3.49	3.03	2.70	2.43	2.21	2.03	1.87
59.5	10.63	67.5	10.36	70	7.93	5.90	4.65	3.86	3.23	2.87	2.55	2.29	2.09	1.92	1.75
59.0	10.00	67.0	10.00	69	7.44	5,51	4.38	3.63	3.10	2.70	2.40	0.16	1.97	1.81	1.66
58.5	9.38	66.5	9.67	68	6.98	5.17	4.11	3.41	2.91	2.54	2.25	2.03	1.85	1.69	1.56
58.0	8.77	66.0	9.30	67	6.51	4.83	3.85	3.19	2.72	2.37	2.11	1.90	1.73	1.58	1.47
57.5	8.17	65.0	8.66	66	6.07	4.51	3.59	2.98	2.54	2.22	1.97	1.77	1.61	1.48	1.37
57.0	**5.57**	**64.0**	**7.95**	**65**	**5.63**	**4.18**	**3.34**	**2.77**	**2.37**	**2.07**	**1.84**	**1.66**	**1.50**	**1.37**	**1.28**
56.5	6.99	63.0	7.31	64	5.21	3.88	3.08	2.56	2.19	1.91	1.71	1.52	1.40	1.28	1.18
56.0	6.41	62.0	6.67	63	4.80	3.57	2.85	2.36	2.02	1.76	1.57	1.41	1.30	1.18	1.09
55.0	5.38	61.0	6.05	62	4.38	3,27	2.61	2.16	1.85	1.61	1.45	1.29	1.20	1.08	1.00
54.0	4.17	60.0	5.46	61	3.96	2.97	2.37	1.97	1.68	1.47	1.30	1.17	1.10	0.99	0.89

(*Contd.*)

Appendix X *(Contd.)*

(1)			(2)	(3)	(4)	(5)	(6)	(7)	(8)	(9)	(10)	(11)	(12)	(13)	(14)
53.0	3.10	...	...	60	3.59	2.68	2.14	1.78	1.52	1.32	1.18	1.06	0.99	0.89	0.82
52.0	**3.04**	**59.0**	**4.85**	**59**	**3.21**	**2.39**	**1.91**	**1.59**	**1.36**	**1.18**	**1.06**	**0.99**	**0.89**	**0.30**	**0.74**
51.0	1.01	56.0	4.28	58	2.85	2.11	1.69	1.40	1.20	1.03	0.94	0.84	0.79	0.71	0.66
50.0	0.00	57.0	3.70	57	2.48	1.33	1.47	1.22	1.04	0.91	0.81	0.73	0.69	0.62	0,57
...	...	56.0	3.15	56	2.07	1.56	1.25	1.04	0.89	0.78	0.69	0.63	0.60	0.54	0.48
49.0	...	55.0	2.60	55	1.73	1.29	1.04	0.86	0.74	0.64	0.57	0.53	0.50	0.45	0.39
48.0	**...**	**54.0**	**2.06**	**54**	**1.38**	**1.0S**	**0.82**	**0.68**	**0.58**	**0.50**	**0.45**	**0.42**	**0.40**	**0.36**	**0.31**
47.0	...	53.0	1.54	53	1.05	0.76	0.61	0.51	0.44	0.37	0.35	0.32	0.30	0.27	0.32
46.0	...	52.0	1.03	52	0.70	0.51	0.41	0.34	0.29	0.24	0.25	0.21	0.20	0.18	0.17
45.0	...	51.0	0.50	51	0.36	0.25	0.20	0.17	0.15	0.12	0.12	0.11	0.10	0.09	0.09
44.0	...	50.0	0.00	50	...	...	...	...	...	....	...	...	...	...	...

APPENDIX-XI

Village-wise Age Composition of Respondents

Name of village	Young Age (<30)		Middle Age (31-45)		Upper Middle Age (46-59)		Old Age (>60)		Total Respondents
	No.	% to Total	No.	% to Total	No.	% to Total	No.	% to Total	
Langasu	0	0.00	8	47.06	6	35.29	3	17.65	**17**
Barhgoan	1	5.26	7	36.84	6	31.58	5	26.32	**19**
Thewa	2	16.67	5	41.67	2	16.67	3	25.00	**12**
Fatehpur	0	0.00	8	38.10	9	42.86	4	19.05	**21**
Rasoolpur	0	0.00	2	8.70	12	52.17	9	39.13	**23**
Shahpur	1	4.17	17	70.83	5	20.83	1	4.17	**24**
Amsaur	1	6.67	6	40.00	6	40.00	2	13.33	**15**
Guleth	0	0.00	7	50.00	3	21.43	4	28.57	**14**
Syunsi	17	28.81	28	47.46	5	8.47	9	15.25	**59**
Bangar	1	6.25	4	25.00	4	25.00	7	43.75	**16**
Sheela mala	8	34.78	2	8.70	5	21.74	8	34.78	**23**
Bhatwari	1	6.25	7	43.75	3	18.75	5	31.25	**16**
Ratuda	2	12.50	4	25.00	4	25.00	6	37.50	**16**
Manzaf	0	0.00	6	37.50	3	18.75	7	43.75	**16**
Maletha	1	6.25	5	31.25	7	43.75	3	18.75	**16**

(Contd.)

Appendix XI (*Contd.*)

Name of village	*Young Age (<30)*		*Middle Age (31-45)*		*Upper Middle Age (46-59)*		*Old Age (>60)*		*Total Respondents*
	No.	*% to Total*	*No.*	*% to Total*	*No.*	*% to Total*	*No.*	*% to Total*	
Mukhwa	0	0.00	5	31.25	6	37.50	5	31.25	**16**
Hitaru	0	0.00	4	26.67	3	20.00	8	53.33	**15**
Kaffar khan	2	11.76	6	35.29	3	17.65	6	35.29	**17**
Majkhali	1	6.25	3	18.75	1	6.25	11	68.75	**16**
Lohar khet	1	6.67	3	20.00	6	40.00	5	33.33	**15**
Aare	3	20.00	5	33.33	3	20.00	4	26.67	**15**
Mudiyani	2	11.11	9	50.00	1	5.56	6	33.33	**18**
Dugri phartiyal	0	0.00	7	43.75	5	31.25	4	25.00	**16**
Dumka bangar	4	28.57	6	42.86	2	14.29	2	14.29	**14**
Gethiya	2	16.67	6	50.00	2	16.67	2	16.67	**12**
Langri	0	0.00	8	50.00	3	18.75	5	31.25	**16**
Kalika	6	37.50	6	37.50	3	18.75	1	6.25	**16**
Devnagar	1	6.67	7	46.67	4	26.67	3	20.00	**15**
Motipur	4	26.67	9	60.00	2	13.33	0	0.00	**15**
Malsa	0	0.00	7	46.67	8	53.33	0	0.00	**15**
TOTAL	**61**	**11.34**	**207**	**38.48**	**132**	**24.54**	**138**	**25.65**	**538**

APPENDIX-XII

Village-wise Educational Status of Respondents

Name of village	Illiterate (0.00)		Less Edu. (<1.50)		Moderately Edu. 1.50-2.50		Highly Edu. (>2.50)		Total Respondents
	No.	% to Total	No.	% to Total	No.	% to Total	No.	% to Total	
Langasu	1	5.88	15	88.24	1	5.88	0	0	17
Barhgoan	5	26.32	10	52.63	3	15.79	1	5.26	19
Thewa	2	16.67	5	41.67	1	8.33	4	33.33	12
Fatehpur	5	23.81	12	54.14	3	14.29	1	4.76	21
Rasoolpur	13	56.52	8	34.78	0	0	2	8.7	23
Shahpur	2	8.33	14	58.33	3	12.5	5	20.83	24
Amsaur	2	13.33	11	73.33	1	6.67	1	6.67	15
Guleth	5	35.71	8	57.14	1	7.14	0	0	14
Syunsi	17	28.81	27	45.76	8	13.56	7	11.86	59
Bangar	0	0	8	50	6	37.5	2	12.5	16
Sheela mala	1	4.35	14	60.87	4	17.39	4	17.39	23
Bhatwari	3	18.75	9	56.25	1	6.25	3	18.75	16
Ratuda	6	37.5	10	62.5	0	0	0	0	16
Manzaf	2	12.5	7	43.75	6	37.5	1	6.25	16
Maletha	4	25	6	37.5	5	31.25	1	6.25	16
Mukhwa	3	18.75	5	31.25	8	50	0	0	16
Hitaru	9	60	5	33.33	0	0	1	6.67	15
Kaffar khan	5	29.41	4	23.53	4	23.53	4	23.53	17
Majkhali	3	18.75	12	75	0	0	1	6.25	16
Lohar khet	2	13.33	10	66.67	2	13.33	1	6.67	15
Aare	0	0	10	66.67	4	26.67	1	6.67	15
Mudiyani	5	27.78	9	50..	3	16.67	1	5.56	18
Dugri Phartiyal	2	12.5	10	62.5	2	12.5	2	12.5	16
Dumka bangar	3	21.43	5	35.71	6	42.86	0	0	14
Gethiya	2	16.67	5	41.67	4	33.33	1	8.33	12
Langri	1	6.25	12	75	2	12.5	1	6.25	16
Kalika	0	0	8	50	5	31.25	3	18.75	16
Devnagar	2	13.33	6	40	4	26.67	3	20	15
Motipur	1	6.67	7	46.67	4	26.67	3	20	15
Malsa	2	13.33	4	26.67	7	46.67	2	13.33	15
Total	**108**	**20.07**	**276**	**51.3**	**98**	**18.22**	**56**	**10.41**	**538**

APPENDIX-XIII

Village-wise Family Size of Respondents

Name of village	*Small (1-4)*		*Medium (5-9)*		*Large (10-14)*		*Very Large (>14)*		*Total Respondents*
	No.	*% to Total*	*No.*	*% to Total*	*No.*	*% to Total*	*No.*	*% to Total*	
Langasu	2	11.76	10	58.82	1	5.88	4	23.53	17
Barhgoan	8	42.11	8	42.11	3	15.79	0	0.00	19
Thewa	2	16.67	8	66.67	2	16.67	0	0.00	12
Fatehpur	4	19.05	13	61.90	4	19.05	0	0.00	21
Rasoolpur	1	4.35	14	60.87	2	8.70	6	26.09	23
Shahpur	3	12.50	17	70.83	2	8.33	2	8.33	24
Amsaur	3	20.00	9	60.00	3	20.00	0	0.00	15
Guleth	2	14.29	12	85.71	0	0.00	0	0.00	14
Syunsi	17	28.81	38	64.41	3	5.08	1	1.69	59
Bangar	3	18.75	12	75.00	1	6.25	0	0.00	16
Sheela mala	12	52.17	11	47.83	0	0.00	0	0.00	23
Bhatwari	1	6.25	12	75.00	2	12.50	1	6.25	16
Ratuda	4	25.00	7	43.75	4	25.00	1	6.25	16
Manzaf	1	6.25	9	56.25	4	25.00	2	12.50	16

(Contd.)

Appendix XIII (*Contd.*)

Name of village	*Small (1-4)*		*Medium (5-9)*		*Large (10-14)*		*Very Large (>14)*		*Total Respondents*
	No.	*% to Total*	*No.*	*% to Total*	*No.*	*% to Total*	*No.*	*% to Total*	
Maletha	2	12.50	11	68.75	2	12.50	1	6.25	16
Mukhwa	4	25.00	7	43.75	3	18.75	2	12.50	16
Hitaru	2	13.33	7	46.67	6	40.00	0	0.00	15
Kaffar khan	2	11.76	13	76.47	2	11.76	0	0.00	17
Majkhali	2	12.50	10	62.50	2	12.50	2	12.50	16
Lohar khet	2	13.33	9	60.00	2	13.33	2	13.33	15
Aare	4	26.67	9	60.00	2	13.33	0	0.00	15
Mudiyani	4	22.22	14	77.78	0	0.00	0	0.00	18
Dugri phartiyal	5	31.25	8	50.00	3	18.75	0	0.00	16
Dumka bangar	2	14.29	11	78.57	1	7.14	0	0.00	14
Gethiya	3	25.00	9	75.00	0	0.00	0	0.00	12
Langri	5	31.25	11	68.75	0	0.00	0	0.00	16
Kalika	4	25.00	9	56.25	2	12.50	1	6.25	16
Devnagar	4	26.67	7	46.67	3	20.00	1	6.67	15
Motipur	1	6.67	10	66.67	1	6.67	3	20.00	15
Malsa	3	20.00	10	66.67	2	13.33	0	0.00	15
Total	**112**	**20.82**	**335**	**62.27**	**62**	**11.52**	**29**	**5.39**	**538**

APPENDIX-XIV

Village-wise Actual Size of Holding

Name of village	Landless (0.00	hec.)	Marginal (<1hec.)	Small (1-2 hec.)		Semi-Medium (2-4 hec.)		Medium (>4 hec.)		Total Respondents
	No.	No.	% to Total	No.	% to Total	No.	% to Total	No.	% to Total	
Langasu	0	17	100.00	0	0.00	0	0.00	0	0.00	17
Barhgoan	0	13	68.42	6	31.58	0	0.00	0	0.00	19
Thewa	0	12	100.00	0	0.00	0	0.00	0	0.00	12
Fatehpur	0	7	33.33	7	33.33	4	4.00	3	14.29	21
Rasoolpur	1	0	0.00	10	43.48	11	11.00	1	4.35	23
Shahpur	0	3	12.50	5	20.83	8	8.00	8	33.33	24
Amsaur	0	13	86.67	1	6.67	1	1.00	0	0.00	15
Guleth	0	9	64.29	5	35.71	0	0.00	0	0.00	14
Syunsi	3	56	94.92	0	0.00	0	0.00	0	0.00	59
Bangar	0	16	100.00	0	0.00	0	0.00	0	0.00	16
Sheela mala	0	0	0.00	1	4.35	8	34.78	14	60.87	23
Bhatwari	0	15	93.75	1	6.25	0	0.00	0	0.00	16
Ratuda	0	13	81.25	3	18.75	0	0.00	0	0.00	16

(Contd.)

Appendix XIV (*Contd.*)

Name of village	*Landless (0.00*	*hec.)*	*Marginal (<1hec.)*	*Small (1-2 hec.)*		*Semi-Medium (2-4 hec.)*		*Medium (>4 hec.)*		*Total Respondents*
	No.	*No.*	*% to Total*	*No.*	*% to Total*	*No.*	*% to Total*	*No.*	*% to Total*	
Manzaf	0	13	81.25	1	6.25	2	2.00	0	0.00	16
Maletha	0	10	62.50	6	37.50	0	0.00	0	0.00	16
Mukhwa	0	16	100.00	0	0.00	0	0.00	0	0.00	16
Hitaru	0	11	73.33	2	13.33	2	2.00	0	0.00	15
Kaffar khan	0	16	94.12	1	5.88	0	0.00	0	0.00	17
Majkhali	0	15	93.75	1	6.25	0	0.00	0	0.00	16
Lohar khet	0	15	100.00	0	0.00	0	0.00	0	0.00	15
Aare	0	14	93.33	1	6.67	0	0.00	0	0.00	15
Mudiyani	0	16	88.89	2	11.11	0	0.00	0	0.00	18
Dugri phartiyal	0	14	87.50	1	6.25	0	0.00	1	6.25	16
Dumka bangar	0	7	50.00	3	21.43	3	3.00	1	7.14	14
Gethiya	0	12	100.00	0	0.00	0	0.00	0	0.00	12
Langri	0	16	100.00	0	0.00	0	0.00	0	0.00	16
Kalika	0	14	87.50	0	0.00	2	2.00	0	0.00	16
Devnagar	0	2	13.33	5	33.33	7	7.00	1	6.67	15
Motipur	1	2	13.33	4	26.67	7	7.00	1	6.67	15
Malsa	2	0	0.00	1	6.67	3	3.00	9	60.00	15
Total	**7**	**367**	**68.22**	**67**	**12.45**	**58**	**10.78**	**39**	**7.25**	**538**

APPENDIX-XV

Village-wise Operational Size of Holdings

Name of village	Marginal (<1 hec.)		Small (1-2 hec.)		Semi-Medium (2-4 hec.)		Medium (>4 hec.)		Total Respondents
	No.	% to Total	No.	% to Total	No.	% to Total	No.	% to Total	
Langasu	17	100.00	0	0.00	0	0.00	0	0.00	17
Barhgoan	14	73.68	5	26.32	0	0.00	0	0.00	19
Thewa	12	100.00	0	0.00	0	0.00	0	0.00	12
Fatehpur	7	33.33	7	33.33	2	9.52	5	23.81	21
Rasoolpur	0	0.00	10	43.48	12	52.17	1	4.35	23
Shahpur	3	12.50	5	20.83	8	33.33	7	29.17	24
Amsaur	14	93.33	0	0.00	1	6.67	0	0.00	15
Guleth	9	64.29	5	35.71	0	0.00	0	0.00	14
Syunsi	59	100.00	0	0.00	0	0.00	0	0.00	59
Bangar	16	100.00	0	0.00	0	0.00	0	0.00	16
Sheela mala	0	0.00	1	4.35	8	34.78	14	60.87	23
Bhatwari	15	93.75	1	6.25	0	0.00	0	0.00	16
Ratuda	13	81.25	3	18.75	0	0.00	0	0.00	16
Manzaf	13	81.25	1	6.25	2	12.50	0	0.00	16
Maletha	12	75.00	4	25.00	0	0.00	0	0.00	16

(Contd.)

Appendix XV *(Contd.)*

Name of village	*Marginal (<1 hec.)*		*Small (1-2 hec.)*		*Semi-Medium (2-4 hec.)*		*Medium (>4 hec.)*		*Total Respondents*
	No.	*% to Total*	*No.*	*% to Total*	*No.*	*% to Total*	*No.*	*% to Total*	
Mukhwa	16	100.00	0	0.00	0	0.00	0	0.00	16
Hitaru	11	73.33	2	13.33	2	13.33	0	0.00	15
Kaffar khan	16	94.12	1	5.88	0	0.00	0	0.00	17
Majkhali	15	93.75	1	6.25	0	0.00	0	0.00	16
Lohar khet	15	100.00	0	0.00	0	0.00	0	0.00	15
Aare	14	93.33	1	6.67	0	0.00	0	0.00	15
Mudiyani	16	88.89	2	11.11	0	0.00	0	0.00	18
Dugri phartiyal	14	87.50	1	6.25	0	0.00	1	6.25	16
Dumka bangar	6	42.86	4	28.57	3	21.43	1	7.14	14
Gethiya	12	100.00	0	0.00	0	0.00	0	0.00	12
Langri	16	100.00	0	0.00	0	0.00	0	0.00	16
Kalika	14	87.50	0	0.00	2	12.50	0	0.00	16
Devnagar	4	25.00	4	25.00	7	43.75	1	6.25	16
Motipur	3	20.00	3	20.00	6	40.00	3	20.00	15
Malsa	2	13.33	0	0.00	1	6.67	12	80.00	15
Total	**378**	**70.26**	**61**	**11.34**	**54**	**10.04**	**45**	**8.36**	**538**

APPENDIX-XVI

Village-wise Level of Innovativeness

Name of village	Less	(<4)	Moderately (4-8)		Highly (>8)		Total Respondents
	No.	% to Total	No.	% to Total	No.	% to Total	
Langasu	17	100.00	0	0.00	0	0.00	17
Barhgoan	19	100.00	0	0.00	0	0.00	19
Thewa	12	100.00	0	0.00	0	0.00	12
Fatehpur	12	57.14	9	42.86	0	0.00	21
Rasoolpur	15	65.22	8	34.78	0	0.00	23
Shahpur	11	45.83	13	54.17	0	0.00	24
Amsaur	15	100.00	0	0.00	0	0.00	15
Guleth	14	100.00	0	0.00	0	0.00	14
Syunsi	59	100.00	0	0.00	0	0.00	59
Bangar	16	100.00	0	0.00	0	0.00	16
Sheela mala	23	100.00	0	0.00	0	0.00	23
Bhatwari	16	100.00	0	0.00	0	0.00	16
Ratuda	16	100.00	0	0.00	0	0.00	16
Manzaf	16	100.00	0	0.00	0	0.00	16
Maletha	16	100.00	0	0.00	0	0.00	16
Mukhwa	16	100.00	0	0.00	0	0.00	16
Hitaru	15	100.00	0	0.00	0	0.00	15
Kaffar khan	17	100.00	0	0.00	0	0.00	17
Majkhali	16	100.00	0	0.00	0	0.00	16
Lohar khet	15	100.00	0	0.00	0	0.00	15
Aare	15	100.00	0	0.00	0	0.00	15
Mudiyani	18	100.00	0	0.00	0	0.00	18
Dugri phartiyal	16	100.00	0	0.00	0	0.00	16
Dumka bangar	14	100.00	0	0.00	0	0.00	14
Gethiya	12	100.00	0	0.00	0	0.00	12
Langri	16	100.00	0	0.00	0	0.00	16
Kalika	16	100.00	0	0.00	0	0.00	16
Devnagar	13	86.67	2	13.33	0	0.00	15
Motipur	12	80.00	3	20.00	0	0.00	15
Malsa	6	40.00	9	60.00	0	0.00	15
Total	**494**	**91.82**	**44**	**8.18**	**0**	**0.00**	**538**

APPENDIX-XVII

Village-wise Patterns of Landuse in Kharif Season

Name of village	*1.Rice*	*2. Madua*	*3.Jhangora*	*4.Rajma*	*5.Pulses*	*6.Sugarcane*
	Area (hec.)	*Area*	*Area*	*Area*	*Area*	*Area*
Langasu	5.22	1.00	0.00	0.00	0.94	0.00
Barhgoan	0.00	0.36	0.04	2.52	0.00	0.00
Thewa	0.20	0.00	0.00	0.00	0.00	0.00
Fatehpur	23.84	0.00	0.00	0.00	0.00	12.88
Rasoolpur	24.88	0.00	0.00	0.00	0.00	28.88
Shahpur	74.80	0.00	0.00	0.00	0.00	77.76
Amsaur	3.84	0.74	0.10	0.00	0.00	0.00
Guleth	3.96	1.90	0.24	0.00	0.00	0.00
Syunsi	9.92	3.04	0.00	2.50	0.00	0.00
Bangar	3.34	2.26	0.00	0.00	0.00	0.00
Sheela mala	1.44	3.80	1.24	0.00	0.00	0.00
Bhatwari	4.50	2.46	1.28	0.00	0.00	0.00
Ratuda	3.60	2.44	0.00	0.00	1.18	0.00
Manzaf	2.50	1.34	0.00	0.00	0.00	0.00
Maletha	4.78	1.30	0.00	0.00	0.00	0.00
Mukhwa	0.12	0.00	0.00	1.20	0.00	0.00
Hitaru	6.50	1.12	0.00	0.00	0.04	0.00
Kaffar khan	3.60	2.52	0.00	0.00	0.28	0.00
Majkhali	0.66	1.68	0.00	0.00	0.00	0.00
Lohar khet	3.20	2.57	0.00	0.00	0.18	0.00
Aare	2.16	0.28	0.00	0.00	0.36	0.00
Mudiyani	2.56	2.20	0.00	0.00	0.40	0.00
Dugri phartiyal	4.34	0.80	0.00	0.00	0.00	0.00
Dumka bangar	5.36	0.00	0.00	0.00	0.00	14.70
Gethiya	1.52	0.00	0.00	0.00	0.00	0.00
Langri	4.80	1.96	0.00	0.00	0.00	0.00
Kalika	5.42	0.10	0.00	0.00	0.00	0.00
Devnagar	26.70	4.00	0.00	0.00	0.00	0.00
Motipur	22.80	0.00	0.00	0.00	0.00	0.40
Malsa	181.22	0.00	0.00	0.00	0.00	93.00
Total	**437.78**	**37.87**	**2.90**	**6.22**	**3.38**	**221.16**

APPENDIX-XVIII

Village-wise Patterns of Landuse in Rabi Season

Name of village	*1.Wheat*	*2.Oilseeds*
	Area (hec.)	*Area*
Langasu	5.30	0.24
Barhgoan	4.00	0.00
Thewa	10.00	0.00
Fatehpur	23.84	0.00
Rasoolpur	24.88	0.00
Shahpur	74.80	0.00
Amsaur	3.44	0.00
Guleth	4.42	0.00
Syunsi	9.92	2.52
Bangar	3.34	0.00
Sheela mala	3.92	1.44
Bhatwari	4.68	0.00
Ratuda	3.80	0.00
Manzaf	2.66	
Maletha	4.44	0.00
Mukhwa	0.00	0.00
Hitaru	6.50	0.00
Kaffar khan	4.44	0.00
Majkhali	3.24	0.20
Lohar khet	2.94	0.00
Aare	2.14	0.16
Mudiyani	4.10	0.00
Dugri phartiyal	4.34	2.46
Dumka bangar	3.56	0.00
Gethiya	2.18	0.00
Langri	5.70	0.92
Kalika	5.42	0.86
Devnagar	25.50	0.00
Motipur	20.80	0.00
Malsa	116.85	0.00
Total	**391.15**	**8.80**

APPENDIX-XIX

Agricultural Diversification

Name of village	Apiculture		Vegetable Farming		Floriculture		Horticulture		Total Respondents
	No.	% to Total	No.	% to Total	No.	% to Total	No.	% to Total	
Langasu	0	0.00	6	35.29	0	0.00	5	29.41	17
Barhgoan	2	10.53	16	84.21	0	0.00	16	84.21	19
Thewa	0	0.00	0	0.00	1	8.33	9	75.00	12
Fatehpur	0	0.00	0	0.00	0	0.00	4	19.05	21
Rasool pur	0	0.00	0	0.00	0	0.00	0	0.00	23
Shali pur	0	0.00	0	0.00	0	0.00	12	50.00	24
Amsaur	0	0.00	7	46.67	0	0.00	4	26.67	15
Guleth	0	0.00	7	50.00	0	0.00	3	21.43	14
Syunsi	0	0.00	59	100.00	0	0.00	3	5.08	59
Bangar	0	0.00	16	100.00	0	0.00	16	100.00	16
Sheela mala	21	91.30	23	100.00	0	0.00	23	100.00	23
Bhatwari	0	0.00	10	62.50	0	0.00	1	6.25	16
Ratuda	0	0.00	4	25.00	0	0.00	3	18.75	16
Manzaf	0	0.00	5	31.25	0	0.00	11	68.75	16
Maletha	0	0.00	4	25.00	0	0.00	0	0.00	16
Mukhwa	0	0.00	2	12.50	0	0.00	14	87.50	16
Hitaru	0	0.00	0	0.00	0	0.00	0	0.00	15
Kaffar khan	0	0.00	9	52.94	0	0.00	6	35.29	17
Majkhali	0	0.00	9	56.25	0	0.00	6	37.50	16
Lohar khet	1	6.67	11	73.33	1	6.67	3	20.00	15
Aare	0	0.00	15	100.00	0	0.00	6	40.00	15
Mudiyani	0	0.00	16	88.89	0	0.00	7	38.89	18
Dugri phartiyal	2	12.50	14	87.50	0	0.00	8	50.00	16
Dumka bangar	0	0.00	0	0.00	0	0.00	0	0.00	14
Gethiya	0	0.00	10	83.33	2	16.67	4	33.33	12
Langri	4	25.00	6	37.50	0	0.00	4	25.00	16
Kalika	3	18.75	8	50.00	0	0.00	13	81.25	16
Shakti farm	0	0.00	0	0.00	0	0.00	4	26.67	15
Motipur	0	0.00	6	40.00	0	0.00	1	6.67	15
Malsa	0	0.00	2	13.33	0	0.00	0	0.00	15

APPENDIX-XXI

Agricultural Diversification in Surveyed Villages (Vegetable Farming)

Name of village	Total Res-pon-dents	No. of Farmer		Commercial		Subsistence		Main Source		Sub. Source	
		No.	% to total Resp.	No.	% to total farmer	No.	% to total farmer	No.	% to commr.	No.	% to commr.
Mukhwa	16	2	12.50	2	100.00	0	0.00	0	0.00	2	100.00
Barhgoan	19	16	84.21	14	87.50	2	12.50	12	85.71	2	14.29
Kalika	16	8	50.00	2	25.00	6	75.00	1	50.00	1	50.00
Bhatwari	16	10	62.50	6	60.00	4	40.00	2	33.33	4	66.67
Lohar Khet	15	11	73.33	5	45.45	6	54.55	0	0.00	5	100.00
Ratuda	16	4	25.00	3	75.00	1	25.00	0	0.00	3	100.00
Langasu	17	6	35.29	2	33.33	4	66.67	0	0.00	2	100.00
Aare	15	15	100.00	3	20.00	12	80.00	0	0.00	3	100.00
Langri	16	6	37.50	1	16.67	5	83.33	0	0.00	1	100.00
Hitaru	15	6	40.00	3	50.00	3	50.00	1	16.67	5	63.33
Manzaf	16	5	31.25	5	100.00	0	0.00	1	20.00	4	80.00
Maletha	16	4	25.00	3	75.00	1	25.00	0	0.00	3	100.00
Guleth	14	7	50.00	2	28.57	5	71.43	0	0.00	2	100.00
Bangar	16	16	100.00	2	12.50	14	87.50	0	0.00	2	100.00
Sheela Mala	23	23	100.00	21	91.30	2	8.70	4	19.05	17	80.95
Syunsi	59	59	100.00	12	20.34	47	79.66	4	33.33	8	66.67
Kaffar Khan	17	9	52.94	1	11.11	8	88.89	0	0.00	1	100.00
Majkhali	16	9	56.25	2	22.22	7	77.78	0	0.00	2	100.00
Dugri Phartiyal	16	14	87.50	9	64.29	5	35.71	0	0.00	9	100.00
Mudiyani	18	16	88.89	3	18.75	13	81.25	1	33.33	2	66.67
Gotia	12	10	83.33	8	80.00	2	20.00	2	25.00	6	75.00
Dumka Bangar	14	5	37.71	2	40.00	3	60.00	2	40.00	3	60.00
Thewa	12	7	58.33	4	57.14	3	42.86	4	57.14	3	42.86
Fatehpur	21	8	38.10	3	37.50	5	62.50	2	25.00	6	75.00
Amsaur	15	7	46.67	2	40.00	5	71.43	0	0.00	2	100.00
Rasoolpur	23	5	21.74	5	55.56	3	60.00	1	20.00	4	80.00
Shahpur	24	9	37.50	1	33.33	4	44.44	3	33.33	6	66.67
Dev Nagar	15	3	20.00	2	66.67	2	66.67	1	33.33	2	66.67
Motipur	15	6	40.00	2	33.33	4	66.67	1	50.00	1	50.00
Malsi	15	2	13.33	2	100.00	0	0.00	2	100.00	0	0.00
Total	**538**	**308**	**57.25**	**132**	**42.86**	**176**	**57.14**	**44**	**33.33**	**111**	**84.09**

(*Contd.*)

Appendix XXI (Contd.)

Name of village	Total Respondents	Area Devote to Cultivation (hec.)	Place of Sell				Avg. Monthly Income	Major Problums
			with in village	% to commr. farmer	out side of	% to commr.		
Mukhwa	16	0.14	1	50.00	1	50.00	1500	
Barhgoan	19	6.49	2	14.29	12	85.71	5000	
Kalika	16	0.28	2	100.00	0	0.00	4500	
Bhatwari	16	0.60	3	50.00	3	50.00	3000	
Lohar Khet	15	0.52	2	40.00	3	60.00	4000	
Ratuda	16	0.20	2	66.67	1	33.33	5000	
Langasu	17	0.32	2	100.00	0	0.00	3000	
Aare	15	1.00	2	66.67	1	33.33	1300	
Langri	16	0.20	0	0.00	1	100.00	6000	
Hitaru	15	0.26	2	66.67	1	33.33	4000	
Manzaf	16	0.54	2	40.00	3	60.00	3500	
Maletha	16	0.18	3	100.00	0	0.00	1200	
Guleth	14	0.99	1	50.00	1	50.00	900	
Bangar	16	1.20	2	100.00	0	0.00	5000	
Sheela Mala	23	65.00	8	38.10	13	61.90	9000	
Syunsi	59	4.80	7	58.33	5	41.67	4000	
Kaffar Khan	17	0.30	1	100.00	0	0.00	400	
Majkhali	16	0.42	1	50.00	1	50.00	1000	
Dugri Phartiyal	16	4.00	2	22.22	7	77.78	9000	
Mudiyani	18	0.98	2	66.67	1	33.33	3500	
Gotia	12	0.84	0	0.00	8	100.00	6000	
Dumka Bangar	14	0.26	2	100.00	0	0.00	800	
Thewa	12	0.35	3	75.00	1	25.00	2000	
Fatehpur	21	0.56	3	100.00	0	0.00	1600	
Amsaur	15	0.20	1	50.00	1	50.00	1200	
Rasoolpur	23	0.24	3	60.00	2	40.00	1500	
Shahpur	24	0.42	1	100.00	0	0.00	2000	
Dev Nagar	15	0.26	2	100.00	0	0.00	1800	
Motipur	15	2.45	1	50.00	1	50.00	10000	
Malsi	15	13.40	0	0.00	2	100.00	8000	
Total	**538**	**107.40**	**47**	**35.61**	**65**	**49.24**	**4645**	

APPENDIX-XXII

Agricultural Diversification in the Surveyed Villages (Horticulture)

Name of village	Total Respondents	No. of Farmer		Commercial		Subsistence		Area Devote to Cultivation (hec.)
		No.	% to total Resp.	No.	% to total farmer	No.	% to total farmer	
Mukhwa	16	14	87.50	8	57.14	6	42.86	2.29
Barhgoan	19	16	84.21	15	93.75	1	6.25	2.99
Kalika	16	13	81.25	7	53.85	6	46.15	1.84
Bhatwari	16	1	6.25	0	0.00	1	100.00	0.04
Lohar Khet	15	3	20.00	1	33.33	2	66.67	0.16
Ratuda	16	3	18.75	1	33.33	2	66.67	1.04
Langasu	17	5	29.41	2	40.00	3	60.00	0.26
Aare	15	6	40.00	6	100.00	0	0.00	0.24
Langri	16	4	25.00	0	0.00	4	100.00	0.10
Hitaru	15	5	33.33	2	40.00	3	60.00	0.50
Manzaf	16	11	68.75	7	63.64	4	36.36	18.20
Maletha	16	6	37.50	2	33.33	4	66.67	1.20
Guleth	14	3	21.43	1	33.33	2	66.67	0.16
Bangar	16	16	100.00	3	18.75	13	81.25	1.00
Sheela Mala	23	23	100.00	21	91.30	2	8.70	108.00
Syunsi	59	3	5.08	1	33.33	2	66.67	0.20
Kaffar Khan	17	6	35.29	0	0.00	6	100.00	0.30
Majkhali	16	6	37.50	2	33.33	4	66.67	0.30
Dugri Phartiyal	16	8	50.00	5	62.50	3	37.50	1.28
Mudiyani	18	7	38.89	5	71.43	2	28.57	0.88
Gotia	12	4	33.33	2	50.00	2	50.00	0.42
Dumka Bangar	14	4	28.57	1	25.00	3	75.00	0.25
Thewa	12	9	75.00	5	55.56	4	44.44	4.90
Fatehpur	21	4	19.05	4	100.00	0	0.00	8.20
Amsaur	15	4	26.67	2	50.00	2	50.00	0.26
Rasoolpur	23	8	34.78	3	37.50	5	62.50	0.26
Shahpur	24	12	50.00	12	100.00	0	0.00	49.00
Dev Nagar	15	4	26.67	1	25.00	3	75.00	2.60
Motipur	15	1	6.67	1	100.00	0	0.00	0.40
Malsi	15	2	13.33	1	50.00	1	50.00	0.45
Total	**538**	**211**	**39.22**	**121**	**57.35**	**90**	**42.65**	**207.72**

(Contd.)

Athewala, A.G., 1966: Some New Methods of Crop-Combination, *Geographical Review of India*, Vol. XXXVIII (1), Kolkata.

Aykroyd, W.R., *et al.*, 1962: *The Nutritive Value of Indian Food and Planning of Satisfactory Diet,* New Delhi.

Aykroyd, W.R., 1956: *The Nutritive Value of Indian Diet and Planning of a Satisfactory Diet*, Madras.

Ayyar, N.P., 1969: "Crop Region of Madhya Pradesh: A Study in Methodology". *Geographical Review of India*, Vol. XXXI, No. 1, p. 17

Baba, S. Baba, S.C. and Sanyal, P., 2009: *Food Security, Poverty and Nutrition Policy Analysis,* Academic Press Pub, Burlington, U.S.A.

Bader, M.D.M., 2010: *Disparities in Neighbourhood Food Environment: Implications : Measurement Strategies*, Vol. 86(4).

Banday, M.T. and Bhakat, M., 2009: 'Raising Livestock' in Masood, S. (Ed.): *Peri-Urban Agriculture - Agriculture by Twenty 20 and Beyond*, A.P.H. Pub. Corp., New Delhi.

Bandooni, S.K., 2004: *Land Resource Management and Development in Hill Areas*, Research India Press, New Delhi.

Banerjee, A. (Ed.), 2004: *Food Security and the Public Distribution System Today*, Kanishka Pub., New Delhi.

Bansil, P.C. and Malhotra, S.P., 2006: *Livestock Economy of India,* CBS Pub. & Dist., New Delhi.

Barun, J.V and Bonillo, E.D. (Eds), 2009: *Globalization of Food and Agriculture and the Poor,* Oxford University Press, London.

Barun, J. *et al.*, 1992: *Improving Food Security of the Poor, Concept Policy and Programme*, International Food Policy Research Institute, Washington.

Basu, D., Kulirani, B.F. and Rahy, B.D. (Eds.), 2006: *Agriculture Food Security Nutrition and Health in North-East India*, Mittal Publications, New Delhi.

Bernard, F.E. and Derrick, J.T., 2007: 'Population Pressure and Human Carrying Capacity in Selected Locations of Machakas and Ketni Districts', in Thakur, B. (Ed.), *Perspectives in Resource Management in Developing Countries*, Concept Publishing Company, New Delhi.

Bhakuni, K., 2000: Identification and Development of Scenic Resources of Kumaon Himalaya, (Unpublished Ph.D. Thesis; Department of Geography, Delhi University).

Bhatia, S. S., 1965: Patterns of Crop Concentration and Diversification in India, *Economic Geography*,Vol. 43.

Bhatia, S.S., 1960: An Index of Crop Diversification, *Professional Geographer*, Vol. XXII.

Bhatia, S.S., 1965: *Patterns of Crop Concentration and Diversification and Development of Mountain Regions with Special Reference to Himachal Pradesh*, M.D. Pub. Pvt. Ltd., New Delhi.

Bhatt, J.P., 2003: Prospective Fisheries Development in Uttaranchal State', in Rawat, M.S.S. (Ed.), *Central Himalaya : Environment and Development*, Department of Geography, H.N.B. Garhwal University, Srinagar.

Bhatt, M.S. (Ed.), 2004: *Poverty and Food Security in India: Problems and Policies*, Aakar Books, New Delhi.

Bhattacharya, B., 2001: Food Security, NAP and NATO, *Yojana*, Vol. 45, No. 1.

Bisht, D.S., 2006: *Poverty, Planning and Developing : A Case Study of Uttarakhand State*, Central Himalayan Institute, Dehradun.

Bisht, D.S., 2001:*Guide to Garhwal and Kumaon Hills., Uttarakhand*, Trishul Pub, Dehradun.

Bora, N.S. and Jalal, D.S., 1995: Status and Sources of Nutrition in the Kumaun Himalayas – A Case Study, *Geographical Review of India*, Vol. 57, No. 1.

Borlang, N.E., 2004: 'Feeding a World of 10 Billion People : Our 21st Century Challenge' in Scanes, C.G. and Miranowski, J.A. (Eds.), *Perspectives in World Food and Agriculture*, Blackwell Pub. Co. Iowa.

Bose, S.P. 1961: Characteristics of Farmers who Adopt Agricultural Practices in Indian Villages, *Rural Sociology*, Vol. 26.

Bruno Dorin and Frederi Lady (Eds.) 2009: *Agriculture and Food in India,* Manohar Publishers and Distributors, New Delhi.

CCS Haryana Agriculture University, 2003: *Emerging Challenges to Haryanas Agricultural Economy Vis-à-vis Diversification in Agriculture,* Hissar.

Census of India, 2001: Uttaranchal-*Administrative Atlas* Directorate of Census

Chakravarty, A.K., 1970: Food Sufficiency Patterns in India; *Geographical Review,* Vol. LX, New York.

Champion, H.G. and Seth, S.K., 1969: *A Revised Survey of the Forest Types of India,* Delhi.

Chand, R., 1997: *Agricultural Diversification and Development of Mountain Region with Special Reference to Himachal Pradesh*, M.D. Pub. Pvt. Ltd., New Delhi.

Chand, R., 1999: *Agricultural Diversification in India : Potential and Prospects*, Mittal Pub., New Delhi.

Chang, S.T., 1999: Global Impact of Edible and Medicinal Mushroom on Human Welfare in 21st Century, *International Journal of Medicinal Mushroom*, Vol. I.

Chatto, M.A. *et al.*, 2009: 'Vegetable Gardening' in Masood, S. (Ed.), *Peri-Urban Agriculture: Agriculture by Twenty 20 and Beyond*, A.P.H. Pub. Corp., New Delhi.

Chaturvedi, P. (Ed.) 1997: *Reconstructing Indian Agriculture, Food Security and Panchayati Raj*, Concept Publishing Co., New Delhi.

Chikara, D.V., 2006: Levels of Agricultural Development in Haryana (Unpublished Ph.D. Thesis, Geography Deptt., Delhi University, Delhi.

Chung, K. *et al.*, 1997: *Identifying the Food Insecure : The Application of Mixed Method Approaches in India*; International Food Policy Research Institute, Washington.

Colin, C.G. and Miranowski, J.A., 2004: *Perspectives in World Food and Agriculture*, Blackwell Pub. Co., Iwo.

Copra, *et al.*, (Eds.), 1998: *Crop Productivity and Sustainability-Shaping the Future*, Oxford and IBH Pub. Co., New Delhi.

Doi, K., 1959: The Industrial Structure of Japanese Prefecture, *Proceeding of International Geographical Union, Regional Conference in Japan*, 1957.

Dorin, B. and Landy, (Eds), 2009: *Agricultural and Food in India*, Manohar Pub. And Distributor, N. Delhi.

FAO 1999: *Food Outlook*, No.4, Rome.

FAO, 1987: *The Status of Food Security in the World*; Rome.

FAO, 1999: *The Status of Food Security in the World*; Rome.

Frankenberger, T.R. and McCarton, 1998: The Household Livelihood Security Concept, *Food Nutrition and Agriculture*, FAO, Rome.

Ganguly, G.B., 2006: 'Agriculture, Food and Public Distribution System in North-East India' in Basu, D., Kulirani, B.F. and Ray, B.D. (Eds.), *Agriculture Food Security: Nutrition and Health in North-East India*, A Mittal Publication, New Delhi.

George, P.S. 1999: Some Reflections on Food Security in India, *Indian Journal of Agricultural Economics*, Vol. 54, and No.(2).

George, P.S., 1994: Food Security in South Asia : Performance and Prospects, *Economic and Political Weekly*, Vol.29, No.15, April 30.

George, P.S., 1994: Public Distribution System, Food Subsidy and Production Insecurities, *Economic and Political Weekly*, Vol. 31, No.39, September 28.

Ghayur, A. and Belt, J., 2004: *Data Base on Medicinal Plants*, CGTs, Kolkata.

Gopal Ji, and Suman Bhakshi, 2005: *Human Development in India*, Ane Books, New Delhi.

Gopalan, C. *et al.*, 2000: *Nutritive Value of Indian Foods*, Indian Council of Medical Research, Hyderabad.

Gopalan, C., 2000: Medium Term Strategy for Meeting the Nutritional Needs of the Population, *Population and Human Development-Meeting Some Critical Needs in the New Century*, Commissioned Papers, Planning Commission.

Gorhuom, D. and Watts, M.J., 1997: *Globalising Food : Agrarian Question and Restructuring.*

Gupta, Mukesh, 2004: *Organic Agriculture Development in India*, A.B.D. Publication, Jaipur.

Hamilton, A., 2003: *Medicinal Plants and Conservation: Issues and Approaches*, International Plant Conservation Unit, U.K.

Hanumantha Rao, C.H. and Radha Krishnan, R, 1997: 'National Food Security-A Policy Perspective for India' in Peters, G.S. (Ed.) *Food Security, Diversification and Resource Management: Refocusing the Role of Agriculture,* Ashgate U.S.A.

Husain, M., 1976: A New Approach of Measuring Agricultural Productivity in Sutlej-Ganga Plains of India, *Geographical Review of India*. Vol. 38 (3) Kolkata.

Indian Agricultural Statistics, Vol. I & II, Directorate of Economics and Statistics, Ministry of Agriculture, Government of India, New Delhi.

International Food Policy Research, Washington.

Jana, B.L., 2006: *Diversification in Agriculture*, Sanjanya Book, Delhi.

Jodha, N.S. *et al.,* (Eds.), 1992: *Sustainable Mountain Agriculture : Perspectives and Issues*, Oxford and IBH, New Delhi.

Jodha, N.S., Banskota, M. and Pratap, T., 1992: *Sustainable Mountain Agriculture: Perspectives and Issues*, Vol. I & II Oxford & IBH New Delhi.

Johnson, B.L.C., 1958: Crop Association Regions in East Pakistan, *Geography*, Vol. XLII.

Johnson, D.L. and Lewiss, 1995: *Land Degradation: Creation and Destruction*. Blackwell, Cambridge.

Johnson, G., 1976: Increased Stability of Grain Supplies in Developing Countries, Optional Carry Overs and Insurance, *World Development*, Vol. 4, No. 12.

Joshi, P. K and Gulati, Ashok, 2007: 'Agricultural Diversification in South Asia: Beyond Food Security' in Joshi, P.K. and Gulati, Ashok, (Eds.),

Agricultural Diversification and Smallholders in South Asia, Academic Foundation, New Delhi.

Joshi, S.C., 2004: *Environment and Development: Geo-ecological Overview*, Gyanodaya Pub., Nainital.

Joshi, S.C., 2004: *Uttaranchal: Environment and Development: A Geo-ecological Overview*, Gyanodaya Prakashan, Nainital.

Kostrowicki, J., 1972: 'Some Methods and Techniques to Determine Crop and Other Landuse Combinations as used in Polish Landuse Studies' in Shafi, M., Anas, M. and Siddiqui, M.F. (Eds.), *Proceeding of Symposium on Landuse in Developing Countries*, Department of Geography, Aligarh.

Kaul, T.N. and Dhar, B.L., 2007: *Biology of Cultivation of Edible Mushroom*, Westville Pub. House, New Delhi.

Kendall, M.G., 1939: The Geographical Distribution of Crop Productivity in England, *Journal of Royal Statistical Society*, Vol. 38.

Khusro, A.M., 1965: Measurement of Productivity at Macro and Micro Levels, *Journal of the Indian Society of Agricultural Statistics*, Vol. 17 (2).

Krishnan, M.S. 1961: *Geology of India and Burma,* Madras.

Lal, S.C. and Sharma, P., 2007: 'Economic Viability of Livestock-Fodder Relationship in the Himalayan Watershed' in Thakur, B., *et al., City, Society and Planning*, Concept, Pub. Co., New Delhi.

Lee, E.S., 1966: A Theory of Migration, *Demography*, Vol. 3.

Mahapatra, S.C., 2005: *Poultry Production in India in the Challenged Global Scenario-Opportunities and Challenges,* Proceedings of XXIII IPSACON.

Mahmood, A., 1977: *Statistical Methods in Geographical Studies*, Rajesh Publications, New Delhi.

Maikhuri, R.M. *et al.,* 2004: 'Agriculture is Uttarakhand: Issues and Management Prospects for Economic Development' in Sati, M.C. and Sati, S.P. (Eds.), *Uttarakhand State: Dimensions of Development*, Indus Pub., New Delhi.

Majeed, A., 1992: "Crop Combinational Analysis: A Review of Methodology", in Mohammad, N. (Ed) *New Dimensions in Agriculture Geography,* Volume. IV. Concept Pub. New Delhi.

Masood, S., 2009: *Peri-Urban Agriculture: Agriculture by Twenty 20 and Beyond*, APH Pub. Corp., New Delhi.

Mehta, G.S., 1999: *Development of Uttarakhand: Issues and Perspectives*, APH Pub. Corp., New Delhi.

Minhas, B.S., 1976: Towards National Food Security, *Indian Journal of Agricultural Economics*, Vol. 31, No. 4.

Misger, F.A., Kumar, A., and Masood, S., 2009: 'Water and Fruit Crops' in Masood, S. (Ed.) *Peri-Urban Agricultural: Agriculture by Twenty 20 and Beyond*, A.H.P. Pub. Corporation, New Delhi.

Mishra, S.R., 2004: 'Public Distribution System in India: Some Critical Issues' in Banerjee, A. (Ed.), *Food Security and the Public Distribution System Today*, Kanishka Publication, New Delhi.

Mohammad, N. (Ed.), 1992: *New Dimensions in Agricultural Geography*, Vol.I-VIII, Concept, New Delhi.

Mohammad, N., 2003: 'Regional Pattern of Food Security in India' in Singh, T.P. (Ed.) *Resource Conservation and Food Security*, Concept. Pub. Co., New Delhi.

Mohammad, N., 1977: The Diffusion of an Agricultural Innovation in Haryana, *The Geographer*, Vol. XXI (1), Aligarh.

Mohammad, N., 1975: Crop Association Regions in Ghaghara-Rapti Doab, *The North Eastern Geographer*, Vol.VII (182).

Mohammad, N., 1988: 'Wasteland in India- Some Issues and Problems', in Sharma, S.C. (Ed.), *Utilization of Wasteland for Sustainable Development*, Concept Publishing Company, New Delhi.

Mohammad, N., 1992: *The Ecology of Agricultural System*, Vol.II, Concept Publishing Company, New Delhi.

Mohammad, N., 2002: Regional Patterns of Food Security in Rural India, *Annals of the National Association of Geographers*, India, and New Delhi.

Mohammad, N., Singh, R.B. and Dutta, A. (Eds.), 2007: *Spatial Information Technology for Natural Resource Management*, Concept Pub. Co., New Delhi.

Mohammad, N., 1978: Impact of Social Factors on Diffusion of Agricultural Innovations in Central Trans-Ghaghara Plan, *National Geographical Journal of India*, Vol. XXIV, Varanasi.

Mohammad, N., 1981: *Perspectives in Agricultural Geography*, Vol. I to V, Concept Pub. Co., New Delhi.

Mohammad, N., 1988: *Caste and Primary Occupation: A Geographical Analysis*, Concept Pub. Co., New Delhi.

Mohammad, N. and Majeed, A., 1979: 'Socio-economic Factors and Diffusion of Agricultural Innovations' in Mohammad, A. (Ed.), *Dynamics of Agricultural Development in India*, Concept Publishing Company, New Delhi.

Mohammad, N., 1992: "Technological Change and Spatial Diffusion Agricultural Innovations in Trans Ghaghara Plain", in Mohammad, N. (Ed) *New Dimensions in Agriculture Geography,* Vol. 6, Concept Pub. New Delhi.

Mohammad, N. and Raza, M., 1992: "Agricultural Meteorological Data Base and Agricultural Productivity" in Mohammad, N (Ed) *New Dimensions in Agriculture Geography*, Volume II, Concept Pub. New Delhi

Mohammad, N. and Mamta, A., 2011: Role of Urban Agriculture in Food Security in Northwestern India, in Singh, K.N and Rana, N.K. (Eds) *Holistic Development: A Geographical Perspectives,* R.K Book, New Delhi.

Mohammad, A., Munir, A. and Rehman, A. (Eds.), 2007: *Fifty Years of Indian Agriculture*, Vol. I & II, Concept Pub. Co., New Delhi.

Mukherjee, A. (Ed.), 2002: International Trade and Food Security in India, *Sustainable Agriculture and Food Security: The Impact of Globalisation*, Sage Publications, New Delhi.

Murlidharan, S., 2001: Upholding the Right to Food, *Frontline,* Vol. 18 (188), New Delhi.

Nag, P. (Ed.), 2010: *Atlas of Agricultural Resources of India*, National Atlas and Thematic Mapping Organization, Kolkata.

Narasaiah, M.L., 2003: *Food Security and Irrigation*, Discovery Pub. House, New Delhi.

Neelofar, L. and Masood, S., 2009: 'Floriculture, Landscaping and Recreation' in Masood, S. (ed.), *Peri-Urban Agriculture : Agriculture by Twenty 20 and Beyond*, A.P.H. Pub. Corp., New Delhi.

Negi, G.C.S., 1994: Agro-forestry System in Central Himalayan Mountains of India, *Agro Forestry Today*.

Nelson, H.J., 1955: A Service Classification of American Cities, *Economic Geography*, Vol. XXXI (3). New Delhi

Pande, P.C., 2007: The Value of Traditional Medicinal and Aromatic Plants of Uttaranchal, *Proceeding of V.C.'s Conference*, Nainital.

Pandey, A.C., Swagata, G. and Nathawat, M.S., 2011: *Evaluating Patterns of Temporal Glaciers in Greater Himalayan Range, J & K.*, Geo-Carto International, London.

Pant, B.R., Joshi, R.C. and Jalal, D.S., 1991: Agricultural landuse and Nutrition in Kotadun, Kumaon Himalayas, *Geographical Review of India,* Vol. 53, and No. 1.

Pant, B.R., 2000: 'Geographical Personality of Uttarakhand' in Sati, M.C. and Sati, M.C. (Eds.) *Uttaranchal State: Dimensions of Development,* Indus Pub. Co. New Delhi.

Pascoe, E.H., 1973: *A Manual of Geology of India and Burma-III,* Govt. of India Publication, Delhi.

Pathak, R.K., 2002: 'Diversification under Different Agro-ecosystem' in Sharma, A.K. (Ed.), *Agricultural Diversification : Problems and Prospects*, I.K. International Pub. House, Pvt. Ltd., New Delhi.

Patra, P., 2006: *Land Degradation in Tropical Highlands*, Rajesh Pub., New Delhi.

Pattanayak, M. and Nayak, B.P., 2005: "Crop Diversification in Orissa: A Spatio-Temporal Analysis", *Eco and WPA*, Delhi.

Pokharel, C., 2006: "Agricultural Impressive Diversification" in *Agricultural Diversification in South Asia*, Kathmandu.

Pownall, L.L., 1953: The Functions of New Zealand Towns, *Annals of the Association of American Geographers*, Vol. 43.

Radha, V., 2009: *Women and Rural Household Food Security*, Serials Publications, New Delhi.

Radhakrishna, R., 1991: Food and Nutrition : Challenge for Policy, *Journal of the Indian Society of Agricultural Statistics*, Vol. 53, No. 3, December.

Rafiuallah, S.M., 1965: A New Approach to Functional Classification of Towns, *The Geographer*, Vol. XXII, Aligarh.

Raj, D., 2005: 'Bio Aesthetic Planning' in Bhattacharjee, S.K. (Ed.), *Advances in Floriculture*, Pointer Pub., Jaipur.

Ramachandran, R., 1969: *Diffusion of Agricultural Innovations in Rural India*, Mysore.

Ramachandran, R., 1992: Correlates of Innovations Among Farmers in the Ganga Plain; in Mohammad, N.(Ed) *Human Perception on Technological Change in Agriculture,* Concept Pub. Co. New Delhi

Randhawa, H.A., 1970: *The Kumaon Himalaya*, Oxford & IBH, New Delhi.

Rangaswamy, P. *et al.,* 1972: *India's Changing Farmers*, Allahabad.

Rawat, A.S., 2002: *Garhwal Himalaya: A Study in Historical Perspectives,* Indus Publishing Co., N. Delhi.

Rawat, M.S.S., 2003: *Central Himalaya: Environment and Development* Department of Geography, H.N.B. Garhwal University, Srinagar.

Raychaudhuri, S.P. *et al.,* 1963: *Soils of India*, New Delhi.

Raza, M., 1992: 'Regional Dimensions of Agricultural Development' in Mohammad, N. (Ed.), *New Dimensions of Agricultural Geography*, Vol. IV, Concept Publishing Company, New Delhi.

Rentlinger, S., 1997: *Food Insecurity: Magnitudes and Remedies,* Washington.

Roy, P.K. and Sharma (Eds.), 2010: *Globalisation and Diversification in India*, Regal Pub., New Delhi.

Ruhela, A., 2008: *Agriculture and Food Security*, Oxford Book Co. Jaipur.

Sarris, A. and Taylor, L., 1976: Central Stock, Food Aid and Food Security, *Food, Nutrition and Agriculture*, Vol. 4, No. 2.

Sastri, C.V.S. and Mukherjee, J.D., 2000: Effect of Light Use Efficiency in Mustard Crop, *Journal of Agro-meteorology* Vol. II (1).

Sati, M.C. and Sati, S.P. (Eds.), 2000: *Uttarakhand State: Dimensions of Development,* Indus Publishing Company, New Delhi.

Saxena, A.K. and Nautiyal, J.C., 1997: Analysing Deforestation: A Dynamic System Approach, *Journal of Sustainable Forestry*, Vol. V, Dehradun.

Scanes, C.G. and Miranowski, J.A. (Eds), 2004: *Perspectives in World Food and Agriculture,* Blackwell Pub. Co., Jawa

Sen, A.K., 1981: *Poverty and Famines: An Essay in Entitlement and Deprivation*, Clarendon Press, Oxford.

Sethi, S. Sharma, R.R. and Masood, S., 2009: 'Value Addition of Fruits', in Masood, S. (Ed.), *Peri-Urban-Agriculture: Agriculture by Twenty 20 and Beyond*, A.P.H. Pub. Corp., New Delhi.

Shafi, M. and Aziz, A. (Ed.), 1989: *Food Systems of the World*, Rawat Pub., Jaipur.

Shafi, M. and Raza, M. (Eds.), 1992: *Forest Ecosystem of the World*, Rawat Pub., Jaipur.

Shafi, M., 1972: Measurement of Agricultural Productivity of the Great Indian Plains, *The Geographer*, Vol. 19(1), Aligarh.

Shafi, M., 2006: *Agricultural Geography*, Pearson Education in South Asia, New Delhi.

Shah, G.L., 1985: 'Some Methodological and Pragmatic Issues in Land Capability Classification in Himalayas', in J.S. Singh (Ed.), *Environmental Regeneration in the Himalaya: Concept & Strategies*, Gyanendra Prakashan, Nainital, pp. 114-129.

Shah, G.L. and Rawat, A.S., 2005: A New Himalayan Entity, *Proceedings of V.C's Conference,* Nainital.

Shah, S.L., 1985: 'Some Methodological and Pragmatic Issues in Land Capability Classification in Himalaya', in J. Singh (ed.), *Environmental Regeneration in the Himalaya: Concept & Strategies,* Gyanendra Prakashan, Nainital.

Sharma, A. and Singh, R.K. (Eds), 2007: *Composite Farming Practice and Economic Development,* Abhijeet Pub, New Delhi.

Sharma, A.K. (Ed), 2002: *Agricultural Diversification: Problems and Prospects*; I.K. International Publishing House Pvt. Ltd, New Delhi,

Shrivastava, M.P. and Singh, J.S., 2008: *Dairy Development in the New Millennium.* Deep and Deep. Pub., N. Delhi.

Siddiqui, M.A.. and Masood, S., 2009: Medicinal and Aromatic Plants' in Masood, S. (Ed.), *Peri-Urban Agriculture : Agriculture by Twenty 20 and Beyond*, A.P.H. Pub., Corp., New Delhi.

Siddiqui, M.F., 1975: Crop-Combination and Specialization in India, *The Geographer*, Vol. XXI, Aligarh.

Singh, B., 1948: *Population and Food Planning in India*, Bombay.

Singh, H. and Akhtar, R., 2007: 'Strategies for Dairy Development in Kashmir Valley: A Geographical Perspectives', in Thakur, B. (Ed.), *City, Society and Planning*, Concept Pub. Co., New Delhi.

Singh, J., 1972: A New Technique of Measuring Agricultural Efficiency in Haryana, India, *The Geographer*, Vol. 19 (1) Aligarh.

Singh, J. and Dhillon, S.S., 1984: *Agricultural Geography*, Tata McGraw Hill Pub., New Delhi.

Singh, J.S. (Ed.), 1985: *Environmental Regeneration in Himalaya : Concept and Strategies*, Gyanendra Prakashan, Nainital.

Singh, J.S., 1985: *Environmental Regeneration in Himalaya*, Gyanendra Parkashan, Nainital.

Singh, L.R., 1965: *The Tarai Region of U.P.*, Allahabad.

Singh, N.P. *et al.*, 2004: *Modern Technology on Vegetable Production*, International Book Distributing Co., Lucknow.

Singh, O.P. 1982: *The Himalaya : Nature, Man, and Culture*, Rajesh Pub., New Delhi.

Singh, R.B., 2008: 'Towards a Food Secure India: Making Hunger a History' in *Science-based Agricultural Transformation Towards Alleviation of Hunger and Poverty in SAARC Countries,* New Delhi.

Singh, R.P., 2007: "Role of Bio-fertilizer in Crop Production' in Sharma, A, Singh, R.K. (Eds.), *Composite Farming Practice and Economic Development*, Abhijeet Pub., New Delhi.

Singh, S., 1948: *Concept and Computation of Consumption Units.*

Singh, S., 2001: Human Development and Food Security, *Yojana*, Vol. 45, No. 8. Food Outlook No. 4.

Singh, S., Singh, B.N. and Pandey, A., 2003: 'Environment, Resources and Development' in Rawat, M.S.S. (Ed.) *Central Himalaya:*

Environment and Development, Department of Geography, HNB Garhwal University, Srinagar.

Singh, V.R., 1970: Land Capability—A Method of Classification, *Professional Geographer*, Vol. 22(1).

Singh, J., 1972: A New Technique of Measuring Agricultural Efficiency in Haryana, India, *The Geographer,* Vol. 19 (1) Aligarh.

Sinha, B.N., 1968: Agricultural Efficiency in India, *The Geographer,* Vol. XV Aligarh.

Sinha, R.R.K., 2009: *Dynamics of Land-Caste Relations in India*, Manak Pub. New Delhi.

Srinivasan, P.V., 2009: 'Agriculture and Food Security', in Ray, Shovan (Ed.), *Oxford Handbook of Agriculture.*

Srivastava, M.P. and Singh, J.S., 2008: *Dairy Development in the New Millennium*, Deep and Deep Pub. New Delhi.

Stamp, L.D., 1958: The Measurement of Land Resources, *The Geographical Review*, Vol. 48(1).

Sud, S., 2000: Population and Food Security; *Yojana.*

Sundaram, K.V., 2001: 'Integrating the Ecosystem Approach into Village and Farm Development', in International Conference on *Sustainable Development and Sustainable Life Styles* Contributory Papers, Bhovigyan Publishers, New Delhi.

Suryanarayana, M.H., 1996: Food Security and Calories Adequacy Across States, *Journal of Indian School of Political Economy*, Vol. 7, No.2.

Swaminathan, M.S., 1993: 'Food Security through Sustainable Agriculture', in *Biotechnological Applications for Food Security in Developing Countries*, Oxford and IBH Pub. Co., New Delhi.

Swaminathan, M.S., 2001: Food Insecurity Atlas of Rural India, *Frontline* Vol. 18, No. 11, May 26-June 8, 2001.

Tewari, R.C., 2006: *Geography of India*, Prayag Pustak Bhawan, Allahabad.

Thakur, B. 2011: *Depletion and Degradation of Land Resources of India*, R.N. Dubey Foundation, Allahabad.

Upreti, B.C, 2004: *Uttaranchal: Issues and Challenges of Development Change,* Kalinga Pub. New Delhi.

Uttarakhand: Forest Statistics (2007-08), Forest Department, Govt. of Uttarakhand, Dehradun.

Vaidyanathan, A., 2010: *Agricultural Growth in India Role of Technology Incentives and Institution,* Oxford University Press, New Delhi

Valdes, A. (ed.), *Food Security for Developing Countries*, Westview Press, Colorado.

Valdiya, K.S., 1987: *Environmental Geology: Indian Contact,* Tata McGraw Hill Pub. Co. Ltd., New Delhi.

Valdiya, K.S., 2001: Physical Divisions of Uttarakhand, *Current Science,* Vol. 81 (10).

Verma, K., 2007: *Role of Urban Agriculture in Life Support System: A Case Study of Alipur Block, Delhi.* Unpublished Dissertation. Department of Geography, Delhi University, Delhi.

Waldia, D.N. and Auden, J.B., 1939: Geology and Structure of Northern Indi*a, Memoirs of Geological Survey of India.* Vol. LXXIII, Delhi.

Warnken, Philip F., 1991: '*Food Security and Agricultural Diversification in Guatemala: Analysis of Interrelationship and Implications for Policy*', Department of Agriculture Economics, University of Missouri.

Weaver, J.C., 1954: "Crop-Combination Regions in the Middle West", *Geographical Review,* Vol. XLIV, No. 2, p. 176.

World Bank Development Report, 1986.

Index